Environmental Monitoring with Arduino

Emily Gertz and
Patrick Di Justo

O'REILLY®
Beijing · Cambridge · Farnham · Köln · Sebastopol · Tokyo

Environmental Monitoring with Arduino

by Emily Gertz and Patrick Di Justo

Printed in the United States of America.

Published by O'Reilly Media, Inc., 1005 Gravenstein Highway North, Sebastopol, CA 95472.

O'Reilly books may be purchased for educational, business, or sales promotional use. Online editions are also available for most titles (*http://my.safaribooksonline.com*). For more information, contact our corporate/institutional sales department: (800) 998-9938 or *corporate@oreilly.com*.

Editors: Shawn Wallace and Brian Jepson
Production Editor: Teresa Elsey
Cover Designer: Mark Paglietti
Interior Designers: Ron Bilodeau and Edie Freedman
Illustrator: Robert Romano

January 2012: First Edition.

Revision History for the First Edition:
January 20, 2012 First release

See *http://oreilly.com/catalog/errata.csp?isbn=9781449310561* for release details.

ISBN: 978-1-449-31056-1
[LSI]
1327090762

To all our nieces and nephews, who we hope will make a more understandable world.

Contents

Preface

This book is all about making the invisible visible.

Each project introduces a particular environmental condition, and then teaches you step by step how to build a small, inexpensive electronic device that can monitor that condition, and communicate back what it finds.

When you start monitoring the environment, something happens: You start to understand the world around you in a new way.

Build a water quality tester, and a beautiful, clear-running stream may become a beautiful clear stream with a high particulate count (see Chapter 6).

Build a gadget to measure temperature and humidity, and you'll see for yourself that "high noon" is not the hottest part of the day; that actually comes around 3 p.m. (see Chapter 8).

Build an electromagnetic field detector, and you'll discover even a quiet room is buzzing with unseen, unheard electrical vibrations (see Chapter 4).

We usually turn environmental monitoring over to the scientific experts at government agencies, universities, and corporations. They come armed with complicated and expensive equipment as well as specialized educations, and occasionally their own institutional agendas.

Since the natural environment is complex, even more so for all the stuff we human beings and our activities have added to the mix, this sort of expertise has an important role in our lives and in our communities. Scientific analysis and expertise are key to creating effective regulations that control the impacts human activities have on the environment and our health.

Monitoring the environment for ourselves, however, pulls the curtain back on what all those experts are doing. Understanding brings knowledge, and with knowledge comes the power to make decisions that can change our lives for the better—from lowering the electric bill, to holding polluters accountable, to helping scientists study the changing climate.

How to Use This Book

We suggest that you build the projects that follow in the given order, since they progress from easier to more complex.

If you already have some experience with Arduino, and want more challenges in making and using these gadgets, look for the "Things to Try" section at the end of each project chapter. We make suggestions for changing the build or the programming that will exercise your skills. We hope you'll come up with your own ideas, too, and tell us about them.

One straightforward way to increase each project's difficulty, once you have built and tested a gadget, is to rebuild it in a more permanent way by soldering the components together. We also offer a few general suggestions for creating enclosures—handy and rugged cases for your gadgets—at the end of this book. You can make enclosures as simply or elaborately as you choose.

Finally: We do our best to describe how to build each gadget as clearly as possible. But as it's almost inevitable that even a "simple" project will frustrate you now and then, here are some tips to keep things fun and interesting:

Break it down
: It may be difficult to get a gadget to work correctly the first time. But don't get discouraged! Most of these gadgets didn't work the first time for us, either. What we've found, and what we think will work for you, is to break every gadget down into separate components, typically input and output components.

Don't skip the preliminaries
: Make sure each component works individually before connecting it with others. If it's working on its own, it will be much more likely to work when combined into a gadget.

Save. Back up. Document.
: When it comes to coding, this is our mantra:

1. Save: Save your code frequently as you work on it.
2. Back up: Always back up your code to at least one location other than your hard drive, such as a peripheral drive, memory card, or flash drive.
3. Document: As you write programming code, include comments (more on this in Chapter 2) that explain what the code does; when you look at your code several days later, you might not remember. As you build a gadget, take notes about what you discover, so that you can refer to them later.

Do these three simple things consistently, and when your computer crashes, your laptop falls out of your bag and onto the concrete, or your cat walks across the keyboard, you will be calm in the knowledge that you always have a copy of your work safely stored somewhere else.

Change only one thing at a time

If you decide to make any changes to the code or the design of these gadgets (and we heartily encourage you to do so), we suggest that you change only one thing at a time, and test it before making another change.

This is important because your change may cause the gadget to stop working. If you've made only one change, it will be easy to undo it and return to a working version of the gadget. This allows you to move ahead with confidence, because you know that any glitch is easy to fix.

Mash it up

We've done our best to design these gadgets in a modular fashion, so that with only a little tweaking, you can swap the the input and output components between them. Want to modify the temperature gadget to output to Pachube rather than a display? Go for it! The hardware should be easy to modify, and changing the code will usually be a simple matter of cutting and pasting from one gadget's code to the next.

Granted, some swaps don't seem to make much sense: it might be rather odd to build a thermometer with an audio output. But give it a try if you want to. Who knows what you'll come up with? Here's our own favorite mashup so far: If you combine the audio output of the electromagnetic field detector with the Geiger counter input, and then tweak the code just right, you can make an old-fashioned click-click-click radiation detector, just like in the movies. So mix and match! Have fun! Be bold!

Ask for help

There is absolutely zero shame in asking for help, and there is less than zero shame in asking for help with an Arduino project. The entire Arduino ecosystem is built on a philosophy of open access to knowledge. Some people may know more about building circuits or writing code than you know. You might know more than someone else.

But to a certain extent, no one is an expert, because no one has made Arduino do everything it can do. Whether online or face-to-face, people will be happy to help you learn if you're respectful, gracious, and willing to share. We guarantee that after weeks of feeling like all you ever do is ask questions, there is nothing like the thrill you'll get the first time you're able to help someone else solve a problem.

Here are some resources for connecting with fellow Arduino users:

- *http://www.arduino.cc*: The online home of Arduino features user forums and more.

- *http://forums.oreilly.com/*: O'Reilly, this book's publisher, has an active Arduino user community.
- *http://hackerspaces.org/wiki/Hackerspaces*: The Hackerspace Wiki is a good place to start looking for face-to-face maker workshops and meetups in your area.

Don't be afraid to experiment
: There's more than one way to put together any device in this book. Don't have a 1 megaohm resistor to use in the EMF detector? Try using a 470K resistor plus a 560K resistor instead. They add up to a bit more than 1 megaohm, but that's OK.

  We know that there are other, perhaps even better ways to build each and every one of the gadgets shown here. We hope you'll find them and let us know about them.The code examples in the following chapters are available for download at GitHub at the official code repository for this book (*https://github.com/ejgertz/EMWA/*). We encourage you to monitor this repository for the latest bugfixed code, as well as extended examples by the author and the rest of the social coding community.

Conventions Used in This Book

The following typographical conventions are used in this book:

Italic
: Indicates new terms, URLs, email addresses, filenames, and file extensions.

`Constant width`
: Used for program listings, as well as within paragraphs to refer to program elements such as variable or function names, databases, data types, environment variables, statements, and keywords.

`Constant width bold`
: Shows commands or other text that should be typed literally by the user.

`Constant width italic`
: Shows text that should be replaced with user-supplied values or by values determined by context.

TIP: This icon signifies a tip, suggestion, or general note.

CAUTION: This icon indicates a warning or caution.

Using Code Examples

This book is here to help you get your job done. In general, you may use the code in this book in your programs and documentation. You do not need to contact us for permission unless you're reproducing a significant portion of the code. For example, writing a program that uses several chunks of code from this book does not require permission. Selling or distributing a CD-ROM of examples from O'Reilly books does require permission. Answering a question by citing this book and quoting example code does not require permission. Incorporating a significant amount of example code from this book into your product's documentation does require permission.

We appreciate, but do not require, attribution. An attribution usually includes the title, author, publisher, and ISBN. For example: "*Environmental Monitoring with Arduino* by Emily Gertz and Patrick Di Justo (O'Reilly). Copyright 2012 Emily Gertz and Patrick Di Justo, 978-1-4493-1056-1."

If you feel your use of code examples falls outside fair use or the permission given above, feel free to contact us at *permissions@oreilly.com*.

Safari® Books Online

Safari Books Online is an on-demand digital library that lets you easily search over 7,500 technology and creative reference books and videos to find the answers you need quickly.

With a subscription, you can read any page and watch any video from our library online. Read books on your cell phone and mobile devices. Access new titles before they are available for print, and get exclusive access to manuscripts in development and post feedback for the authors. Copy and paste code samples, organize your favorites, download chapters, bookmark key sections, create notes, print out pages, and benefit from tons of other time-saving features.

O'Reilly Media has uploaded this book to the Safari Books Online service. To have full digital access to this book and others on similar topics from O'Reilly and other publishers, sign up for free at *http://my.safaribooksonline.com*.

How to Contact Us

Please address comments and questions concerning this book to the publisher:

O'Reilly Media, Inc.
1005 Gravenstein Highway North
Sebastopol, CA 95472
800-998-9938 (in the United States or Canada)
707-829-0515 (international or local)
707-829-0104 (fax)

We have a web page for this book, where we list errata, examples, and any additional information. You can access this page at:

http://shop.oreilly.com/product/0636920021582.do

To comment or ask technical questions about this book, send email to:

bookquestions@oreilly.com

For more information about our books, courses, conferences, and news, see our website at *http://www.oreilly.com*.

Find us on Facebook: *http://facebook.com/oreilly*

Follow us on Twitter: *http://twitter.com/oreillymedia*

Watch us on YouTube: *http://www.youtube.com/oreillymedia*

1/The World's Shortest Electronics Primer

If you're a DIY electronics or Arduino novice, the information in this chapter will help you get the most out of building and programming the gadgets in this book.

If you're already building your own electronics, consider this chapter a refresher to dip into as needed.

What Is Arduino?

Arduino is best described as a single-board computer that has deliberately been designed to be used by people who are not experts in electronics, engineering, or programming. It is inexpensive, cross-platform (the Arduino software runs on Windows, Mac OS X, and Linux), and easy to program. Both Arduino hardware and software are open source and extensible.

Arduino is also powerful: despite its compact size, it has about as much computing muscle as one of the original navigation computers from the Apollo program, at about 1/35,000 the price.

Programmers, designers, do-it-yourselfers, and artists around the world take advantage of Arduino's power and simplicity to create all sorts of innovative devices, including interactive sensors, artwork, and toys.

We built each of the products in this book using the Arduino Uno (Figure 1-1 and Figure 1-2), which at this writing (late 2011) is the latest model. By the time you're reading this, there may be something newer.

It's not necessary to know Arduino Uno's technical specifications to build and program the gadgets in this book. But if you're interested, you can find them at the official Arduino website (*http://arduino.cc/en/Main/ArduinoBoardUno*).

Electronic Circuits and Components

An electronic circuit is, as the term implies, electricity moving in a path very much like a circle. Each circuit has a beginning, a middle, and an end (which

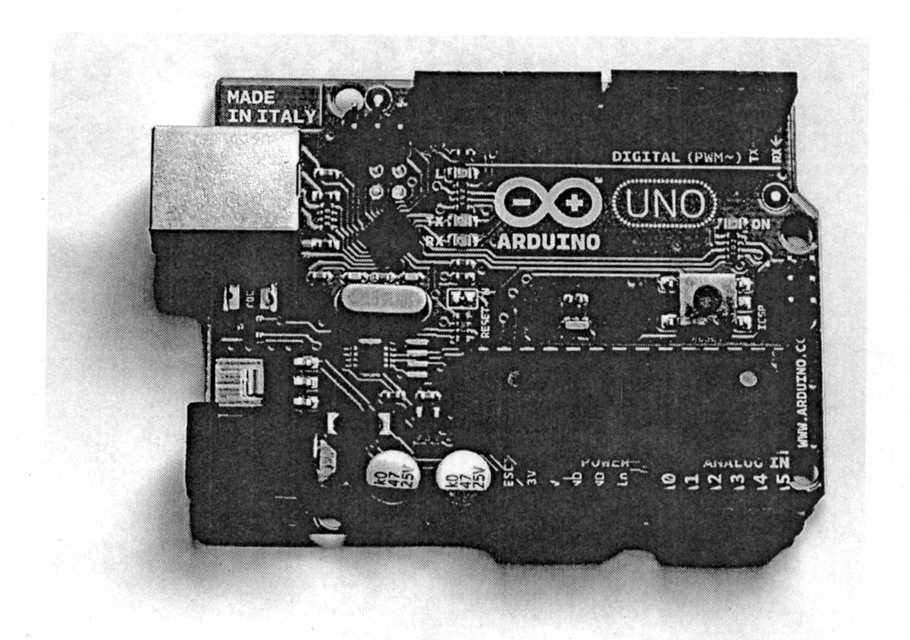

Figure 1-1. *Front of the Arduino Uno (Rev. 2).*

is usually very close to where it began). Somewhere in the middle, the circuit often runs through various electronic components that modify the electrical current in some way.

Each device in this book is a circuit that combines Arduino with different electronic components. Some of these essentially manage the power and path of the electricity; others sense certain conditions in the environment; and still others display output about those conditions.

Let's take a look at some of the components we will be using in our circuits:

Light emitting diodes (LEDs)

An LED is a lamp made of various rare-earth metals, which give off a large amount of light when a tiny current is run through them. The composition of the substances within the LED determine the particular wavelength of light emitted: green, blue, yellow, red, and even ultraviolet and infrared are among the possible colors.

Technically, the LEDs used in our gadgets are "miniature LEDs," tiny lamps with two wire leads, one long (called the anode) and the other a bit shorter (called the cathode). These come in various useful forms, including single lamps from 2mm to 8mm in diameter, display bars, and

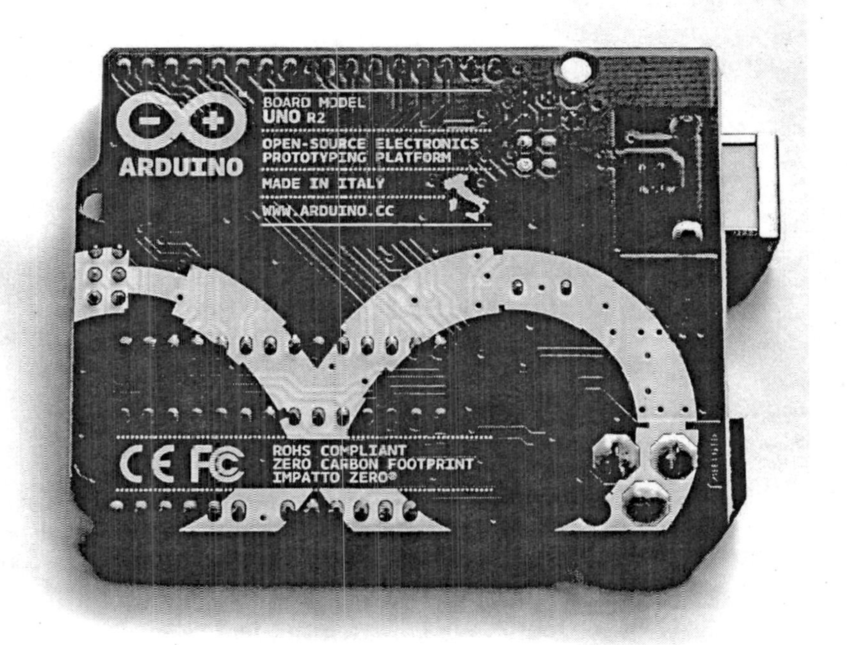

Figure 1-2. *Back of the Arduino Uno.*

alphanumeric readouts, and can serve as indicators, illuminators, or even data transmitters.

You'll learn how to use these different types of LEDs while building the different environmental sensors in this book.

Resistors

Resistors are the workhorses of the electronics world. What do resistors do? They simply resist letting electricity flow through them, and they do this by being made of materials that naturally conduct electricity poorly. In this way resistors serve as small dumb regulators to cut down the intensity of electric current.

Resistance is valuable because some electronic components are very delicate, burning out easily if they're powered with too much current. Putting a resistor in the circuit ensures that only the proper amount of electricity reaches the component. It's hard to imagine any circuit working without a resistor, and with LEDs resistors are almost mandatory.

While building the projects in this book, you'll learn various creative ways to regulate current with resistors.

Soldering

Soldering involves heating up conductive metal, called solder, and then using it to fuse other pieces of metal together. In small-scale electronics, we use an electrical tool called a soldering gun, which has a small tip, to heat up thin wires of solder and drip the solder onto the components we wish to join into the circuit.

Soldering creates a very stable circuit, and that stability can be a drawback. Fusing together components can make it difficult to reuse or reconfigure circuits. You also must be very careful to not short-circuit components while soldering. It is beyond the scope of this book to to go into the details of soldering, which can be a very useful skill in DIY electronics. If you're interested in learning how, this online resource (*http://mightyohm.com/files/soldercomic/FullSolderComic_20110409.pdf*) is a good place to start.

The alternative to soldering is to use a breadboard.

Solderless breadboards

Solderless breadboards are small plastic boards studded with pins that can hold wires. (More about these below.) These wires can then be connected to other electronic components, including Arduino.

Solderless breadboards make it much easier to design circuits, because they allow you to quickly try out various assemblies and components without having to solder the pieces together. While solderless breadboards typically are intended for use only in the design phase, many hobbyists keep a breadboard in the final version of a device because they're so fast and easy to use.

If you don't feel like soldering circuit boards, solderless breadboards are the way to go. Each gadget in this book uses a solderless breadboard.

Wire

Wire is the most basic electronic component, creating the path along which electrons move through a circuit. The projects in this book use 1mm "jumper wires," which have solid metal tips perfectly sized to fit into Arduino and breadboard pins, and come sheathed in various colors of insulation.

Get as much jumper wire as you can afford, in several colors. When building circuits with Arduino, you can't have too many jumper wires.

We order most of our electronics components from these online retailers:

- Adafruit Industries (*http://adafruit.com*)
- Eemartee (*http://www.emartee.com*)
- Electronic Goldmine (*http://www.goldmine-elec.com*)
- SparkFun (*http://www.sparkfun.com*)

Maker Shed, from MAKE and O'Reilly Media, sells books, kits, and tools, as well as many of the electronic components needed to build the projects in this book. Maker Shed also supplies convenient bundles for many of the projects in this book (you can find more information about these bundles in the individual project chapters).

Don't count out your friendly local RadioShack (*http://www.radioshack.com*), though. While writing this book, more than once we ran out to RadioShack for a last-minute component.

For years RadioShack cut back on its electronic components inventory, apparently seeing a better future for the business by featuring cell phones and other consumer electronics. But the company has recently begun to embrace the maker movement; at this writing, some stores around the country are even carrying Arduinos. We're hopeful RadioShack is on the return path to being the hacker heaven it was years ago.

Programming Arduino

A computer program is a coded series of instructions that tells the computer what to do. The programs that run on Arduino are called *sketches*.

The sketches used in this book mostly tell Arduino to read data from one of the pins, such as the one connected to a sensor; and to write information to a different pin, such as the pin connected to an LED or display unit.

Sometimes the sketches also instruct Arduino to process that information in a certain way: to combine data streams, or compare the input with some reference, or even place the data into a readable format.

An Arduino program has two parts: `setup()` and `loop()`.

setup()

The `setup()` part tells Arduino what it needs to know in order to do what we want it to do. For example, `setup()` tells Arduino which pins it needs to configure as input, which pins to configure as output, and (by default) which won't be doing anything. If we're going to use a special type of output to show our results, such as a four-character display, `setup()` is where we tell Arduino how that output works. If we need to communicate with the outside world through a serial port or an Ethernet connection, all the instructions necessary to make that connection go here.

loop()
: `loop()` tells Arduino what to do with the input or output. Unlike some other computers, it never stops; once the instructions in a loop have been executed, Arduino goes right back to the top of the `loop()` and starts executing instructions all over again.

First Sketch: Make an LED Blink

By long tradition (going back to 2006), the first Arduino sketch you write is to make an LED blink.

Arduino pins can be used for input and output, as long as you tell the computer which is which. So in this sketch, we tell the Arduino to set pin 13 to be the LED OUTPUT pin, and then we alternately send electricity to pin 13 (setting the pin HIGH) and cut off the electricity to pin 13 (setting the pin LOW). With each alternation, the LED turns on and off.

We'll write all the sketches in this book using the Arduino "integrated development environment" (IDE), which, simply put, is special software for writing and uploading code to Arduino.

Parts

1. Arduino Uno
2. Breadboard
3. LED

Install the IDE

Download the Arduino IDE from *http://arduino.cc/en/Main/Software*, and follow the provided instructions to install it on your computer.

Once you've installed the software, open the IDE. You should see a screen that looks something like Figure 1-3.

Breadboard the Circuit

The circuit portion of this project is very simple:

Take an LED and place the long lead into pin 13 on Arduino, as you can see in the Figure 1-4 breadboard view.

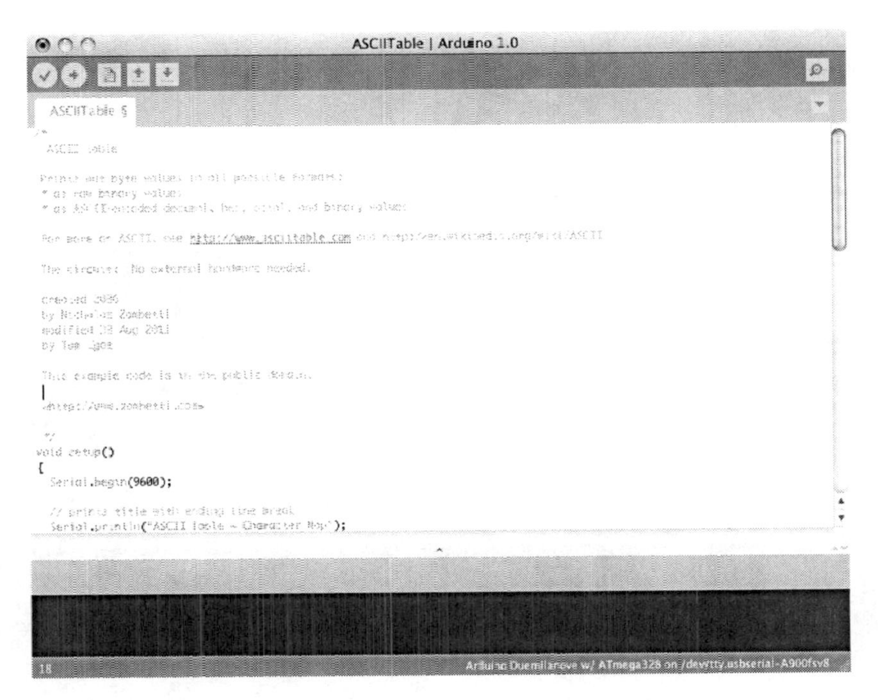

Figure 1-3. *The Arduino IDE on a Mac.*

Write the Code

You can find this code in the Arduino IDE under File → Examples or on the EMWA GitHub Repository | chapter-1 | blink (*https://github.com/ejgertz/EMWA/blob/master/chapter-1/blink*).

```
/*
  Blink
  Turns on an LED on for one second,
  then off for one second, repeatedly.
  This example code is based on example code
  that is in the public domain.
*/

void setup() {
  // initialize the digital pin as an output.
  // Pin 13 has an LED connected on most Arduino boards:
  pinMode(13, OUTPUT);
}

void loop() {
  digitalWrite(13, HIGH);   // set the LED on
  delay(1000);              // wait for a second
  digitalWrite(13, LOW);    // set the LED off
```

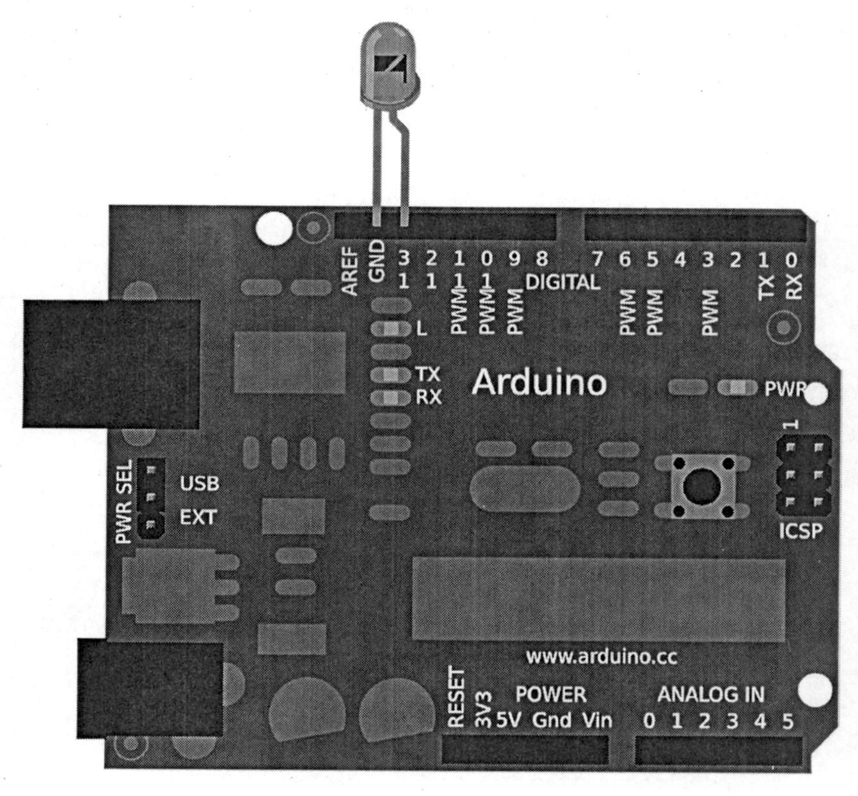

Figure 1-4. *LED long lead inserted into pin 13 on the Arduino (image made with Fritzing.org).*

```
    delay(1000);              // wait for a second
}
```

In this sketch, the code in `loop()` simply tells Arduino to set pin 13 HIGH—taking it up to 5 volts—for 1000 milliseconds (one second), followed by setting it LOW—taking it down to 0 volts—for another 1000 milliseconds.

Notice the /* ... */ sections and the // lines in the example above? Those are ways to put comments into your code to explain to others (and to yourself) what the code does: /* and */ tell the computer that everything between those marks should be ignored while running the program. // tells the computer that everything afterward on that line is a comment.

Why Comment Code?

Commenting code simply means adding explanations in plain English to your sketch that describe how the code works. Adding comments to code is a very good idea. Here's why:

Suppose, after hours trying to get your Arduino to do something, the solution suddenly comes to you. Eureka! You hook up your Arduino, bang out your code, load it up, and voilà: It works.

Fast forward: Months later, working on another project, you want your Arduino to do something similar to your earlier project. "No sweat, I'll just reuse my earlier code," you think. But you open up the sketch and...none of it makes sense!

You wrote that earlier code in a highly creative state of mind, when your brain chemicals were flowing like a river and your ideas were flashing like summer lightning. In all the excitement, you didn't comment your code. So now, months later, when you're in a completely different state of mind, you can't remember what the code does, and you have to start all over Is that any way to live?

If you had commented your code from the beginning, you'd know exactly what each variable was used for, what each function did, and what each pin controlled. Your life would be so much more enjoyable.

In short, always take a few minutes to comment your code.

Things to Try

Modify this sketch to make the LED do something different:

1. Blink twice as quickly.
2. Blink twice as slowly.
3. Light up for half a second with a two-second pause between blinks.

Congratulations, you're an Arduino programmer! Now let's have some real fun.

2/Project: Noise Monitor/ LED Bar Output

We cannot smell, taste, or touch a sound. But noise (which is what most of us call a sound we don't like) is one of the most pervasive environmental contaminants around.

Noise pollution is defined as a sound that is constant, very loud, unwanted, or disturbing to everyday activities in the places we live, play, work, or learn. Cars on the street, planes overhead, construction equipment, or your neighbor's loud TV leaking through the wall—these and more can become noise pollution. And it's not merely a case of acute annoyance: According to the U.S. Environmental Protection Agency, noise pollution is directly linked to stress and stress-related illnesses ("all that noise is making me sick"), high blood pressure, fatigue, and hearing loss, among many other adverse effects.

Even the thick-skinned residents of New York City lose their cool when it comes to noxious sounds: unwanted noise is far and away the number-one complaint to the city's 311 info and services line.

Measuring Noise: The Microphone

Sound is made by the movement of air molecules. When an object vibrates, it moves back and forth, creating pressure waves that compress the air first in one direction, and then in the other. These waves of compression travel outward in all directions from the source of the vibration until they hit an obstacle and get absorbed, reflected, or attenuated into nothingness.

When the wave reaches our microphone, its pressure causes a membrane in our microphone to vibrate. As the microphone membrane vibrates, it changes the magnetic field of a magnet behind it. This varying magnetic field causes a very small electric current to flow from the microphone's wires. That current is what we actually measure with this gadget.

Typically a microphone current is very low—so low that Arduino would find it difficult to detect much variation in the signal. So we chose the Mini Sound Sensor mic (Emartee part number 42021) (*http://www.emartee.com/product/41496/Mini%20Microphone*). This mic comes loaded onto a breakout board equipped with an amplifier. This particular amp boosts the signal to one strong enough for Arduino to detect easily, which gives us a lot to work with.

If the Emartee Mini Sound Sensor isn't available when you're reading this book, a mini microphone from Jameco (part number ECM-60PC-R) (*http://www.jameco.com/webapp/wcs/stores/servlet/Product_10001_10001_136574_-1*) should also work, although it may require some tweaking of the Arduino sketch for this gadget.

Save the Whales...from Noise Pollution

We've been talking about pressure waves moving through the air, but noise can move just as easily through nearly any continuous medium: metals, glass, even water. In fact, there is a growing body of proof that increasing levels of undersea noise, largely caused by ship engines, are harming social sea mammals like dolphins and whales.

These animals, which communicate using underwater sound, are having a harder time talking to one another because of all these unmuffled engines. Scientists using underwater recording devices published research in 2010 (*http://www.onearth.org/blog/right-whales-yell-over-noise-pollution*) showing that endangered North Atlantic right whales are being forced to turn up their call volume to find each other over the undersea din. If they can't find each other, they can't mate and produce offspring.

Modifying this gadget to listen to ocean noise would make a great project, albeit a complicated one The microphone would need to be waterproofed, as well as designed to pick up the frequencies used by creatures like dolphins and whales. A waterproof housing would be essential for Arduino itself as well, plus a method to either store the data (see more about the SD card in "The SD Card Slot" on page 50) or output the data to a device elsewhere.

If you attempt this, remember to let us know how it turns out!

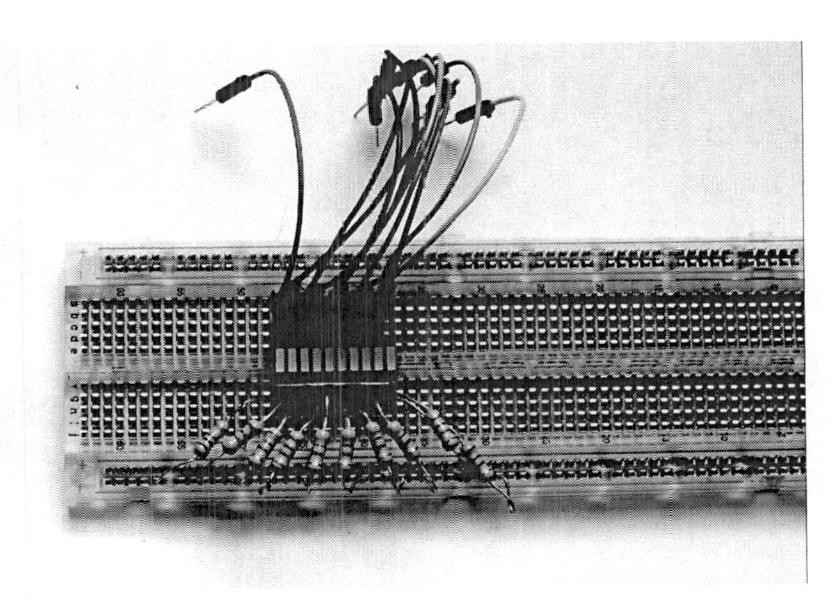

Figure 2-1. *An LED bar display plugged into a breadboard, along with jumper wires to connect it to Arduino. You can also create an LED bar display using individual LEDs, as seen in the breadboard view.*

The LED Bar

The LED bar display, available from SparkFun (sku COM-09935) (*http://www.sparkfun.com/products/9935*) and other electronics suppliers, is nothing but a collection of light emitting diodes in a fancy plastic case (see Figure 2-1). There is no other circuitry. There aren't even any built-in resistors to regulate the current. For that reason, we stress strongly that if you do not want to use the LED bar, you certainly don't have to.

Feel free to substitute any number of standard LEDs in its place. Just be certain to change the variable `number_of_LEDs` in the sketch to reflect the actual number of LEDs that you use.

One advantage to using individual LEDs is that you can color-code them by intensity. Try five green LEDs, three yellow LEDs and two red LEDs to give your readout a sense of urgency.

Make the Gadget

Parts

1. Arduino
2. Breadboard
3. Mini Sound Sensor microphone (Emartee part number 42021)
4. 5–10 LEDs, one or more colors, or LED bar display
5. 220-ohm resistor
6. 10–15 jumper wires in varied colors

Breadboard the Circuit

You can see what the final build looks like in the breadboard view of this circuit in Figure 2-2.

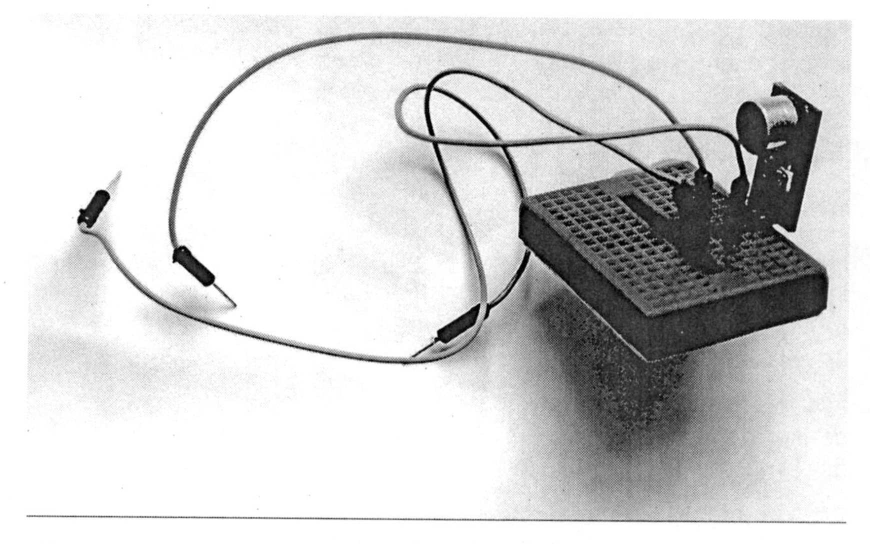

Figure 2-2. *The completed noise monitor circuit.*

Here's how to build that circuit:

Step 1 Plug the microphone into the breadboard (see Figure 2-3).

Step 2 Connect a wire between the GND pin of the microphone and the GND pin of Arduino.

Step 3 Connect the power pin of the microphone to the power pin of Arduino.

Step 4 Connect the DATA pin of the microphone to the Analog 0 pin of Arduino.

Step 5 Connect the Digital 2 pin of Arduino to a point on the breadboard.

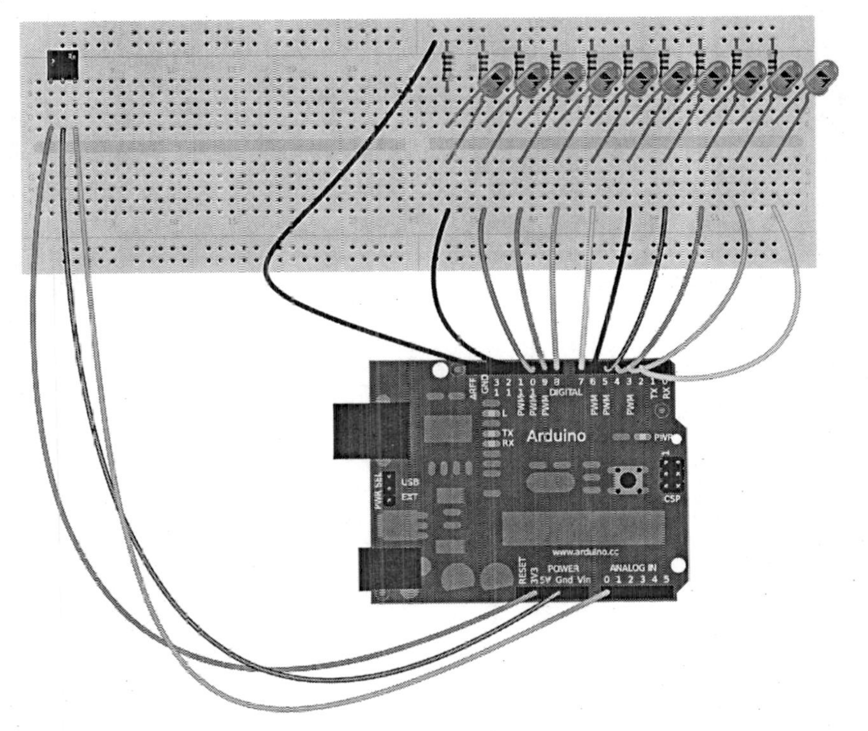

Figure 2-3. *The noise sensor plugged to the breadboard, with jumper wires leading from its GND, power, and DATA pins.*

Step 6 Connect the LONG or ANODE lead of an LED (or the ANODE lead of an LED bar) to a pin in the same breadboard row as the jumper from D2. Have the LED straddle the breadboard trench, and plug the SHORT lead or CATHODE (or the CATHODE lead of an LED bar) to a pin in the corresponding row on the other side of the breadboard.

Step 7 Plug a 220-ohm resistor into the breadboard, connecting the cathode row and the GND rail.

Step 8 Connect a wire from the GND rail to the Arduino GND pin.

Repeat steps 5 through 7 nine times—or once for every LED you want to use. Increase the digital Arduino pin and breadboard row for each LED, to make a nice row of lights.

To keep yourself from going crazy, don't use the same color wire for each LED, since that makes it unbelievably difficult to spot mistakes made by plugging an LED to the wrong Arduino pin. Alternate colors, or use a whole rainbow of wires.

Write the Code

You can find this sketch on the EMWA GitHub repository | chapter-2 | NoiseMonitor (*https://github.com/ejgertz/EMWA/blob/master/chapter-2/NoiseMonitor*).

```
/*
  Noise Monitor
  Sketch for an Arduino gadget that detects noise.
  This example code is based on example code that is in the public domain.
 */

int sensorPin = A0; // select the input pin for the input device

const int numberOfLEDs = 10;

const int numberOfSamples = 16;
int sample;
long signal[numberOfSamples];
long runningAverage;
long sumOfSamples = 0;
int counter =0;

int threshold[] = { 0, 47, 99, 159, 227, 308, 407, 535, 715, 800, 900};

// You can play with the sensitivity of the LEDs by removing the above
// threshold and using the one below. Try different values. Experiment!
//int threshold[]={ 0, 25, 50, 75, 100, 125, 150, 175, 200, 225};

void setup()
{
  // declare the ledPins as an OUTPUT.
  // We're doing it line-by-line, so you can see what's happening.
  pinMode(2, OUTPUT);
  pinMode(3, OUTPUT);
  pinMode(4, OUTPUT);
  pinMode(5, OUTPUT);
  pinMode(6, OUTPUT);
  pinMode(7, OUTPUT);
  pinMode(8, OUTPUT);
  pinMode(9, OUTPUT);
  pinMode(10, OUTPUT);
  pinMode(11, OUTPUT);
```

```
  // setting each pin to LOW so as not to light the LED
  digitalWrite(2, LOW);
  digitalWrite(3, LOW);
  digitalWrite(4, LOW);
  digitalWrite(5, LOW);
  digitalWrite(6, LOW);
  digitalWrite(7, LOW);
  digitalWrite(8, LOW);
  digitalWrite(9, LOW);
  digitalWrite(10, LOW);
  digitalWrite(11, LOW);

  // set the analog 0 pin to input
  pinMode(sensorPin, INPUT);

  // Getting a baseline noise signal
  for(int i =0; i <=numberOfSamples; i++)
  {
    sample = analogRead(sensorPin);

    signal[i] = abs(sample -512);
    sumOfSamples = sumOfSamples + signal[i];
  }

  // Tests the LEDs by turning them on.
  // This time, we're using a for() loop to do the job.
  // Using for(), while(), and other loops is probably
  // how you should handle tasks like this.
  for(int i=0; i <=numberOfLEDs; i++)
  {
    digitalWrite(i+1, HIGH);
    delay(100);
  }

  // ... and then turning them off.
  for(int i=0; i <=numberOfLEDs; i++)
  {
    digitalWrite(i+1, LOW);
    delay(100);
  }

  Serial.begin(9600);
}

void loop()
{
  // We want to take a "running average" of the output of the
  // microphone.  We started getting a baseline average back
  // in setup().  Now, we're subtracting the oldest sound
  // sample from the running total, taking a new sound sample,
  // adding that to the running total, and taking the average.
```

```
  // This gives us a "typical" sound sample.

  // Here we increase our counter, to keep track of how many
  // audio samples we're taking.  If we use more than the number
  // of samples, use the % (modulo) operator to set the counter to zero.

  counter = ++counter % numberOfSamples;

  // subtract the oldest sample from our total audio sample
  sumOfSamples -= signal[counter];

  // take a new audio sample
  sample = analogRead(sensorPin);

  // assign the sample to an array, normalize and adjust it to remove
  // negative values
  signal[counter] = abs(sample -512);

  // Add the most recent sample to the total audio sample
  sumOfSamples = sumOfSamples + signal[counter];

  // And (this is the key part), take an average of all the samples
  runningAverage = sumOfSamples/numberOfSamples;

  Serial.print("Running Value = ");
  Serial.println(runningAverage);
  Serial.println(" ");

  // light up the LEDs
  for (int i =0; i <=numberOfLEDs; i++)
  {
    // Then see if the average sound meets that LED's threshold value
    if(runningAverage>threshold[i])
    {
      // if so, light the LED
      digitalWrite(i+1, HIGH);
      delay(10);
    }
  }

  // turn all LEDs off from right to left.  This keeps the display
  // "active", like the display on an audio amplifier
  for (int i =numberOfLEDs; i >=1; i--)
  {
    digitalWrite(i+1, LOW);
  }
}
```

Things to Try

1. Adapt the device for underwater listening, as suggested earlier.
2. Adjust the code so that the LEDs display the loudest noise on a sliding, not fixed, scale.
3. Leave the "maximum" LED lit for a few seconds.

3/New Component: 4Char Display

In the next project, we're going to display our data on a serially driven four-character LED display (see Figure 3-1). This is a wonderfully versatile little tool that incorporates four seven-segment LED displays that show, natch, four characters of data at a time (as well as a colon and decimal points).

The four-character display can show all of the Arabic numerals from 0 to 9, as well as 20 of the 26 letters used in English, some of them in both upper and lower cases. To see how they look, check out Figure 3-2. (There are some pseudoconventions for displaying the letters k, m, v, w, and x, but if you use them, most people won't recognize them as letters—some of them look just like random illuminated segments—and will think there's something screwy with the display).

"Serially driven" is the distinction that makes this display so useful. With a standard seven-segment display, each segment of the display needs its own dedicated data line from Arduino to control it. Using four characters in this type of display architecture (along with the associated decimal points and colon) would require 34 dedicated data lines, more than the standard Arduino even has. (Granted, there are tricks to get around this, but even then, the display would still need a lot of lines.)

So the people at SparkFun, who make this product, added a microcontroller to the back of the display. This microcontroller can take serial data sent from a single Arduino pin and interpret it to properly control all four display characters.

But every boon has a price. In this case, the boon is that we need only two dedicated data lines to use the 4Char; the price is that you must format your data so that it is sent in groups of four characters at a time. Always. No exceptions. Do you have only three characters to display? Too bad. You must add a space or a legend character so that you're feeding exactly four characters to the display. If you have five or more characters to display, you must format your data into four-character chunks and add some code to make your data scroll past the display.

Figure 3-1. *Front (left) and back (right) of the 4Char LED display.*

Figure 3-2. *How English letters look on the seven-segment, four-character display.*

Test Project

To get a feel for how the 4Char display works, let's wire it up and run some sample code.

Parts

1. Arduino
2. Breadboard
3. 4Char display (SparkFun sku COM-09765)+
4. Jumper wires in various colors

Breadboard the Circuit

This is a very simple circuit to build, as you can see in the breadboard view Figure 3-3.

Step 1 Connect a jumper from Arduino GND to the GND pin on the 4Char.

Step 2 Place the jumper through the GND hole on the display, and anchor it in the breadboard. (It doesn't matter which breadboard row you use. Try not to use one of the rails.)

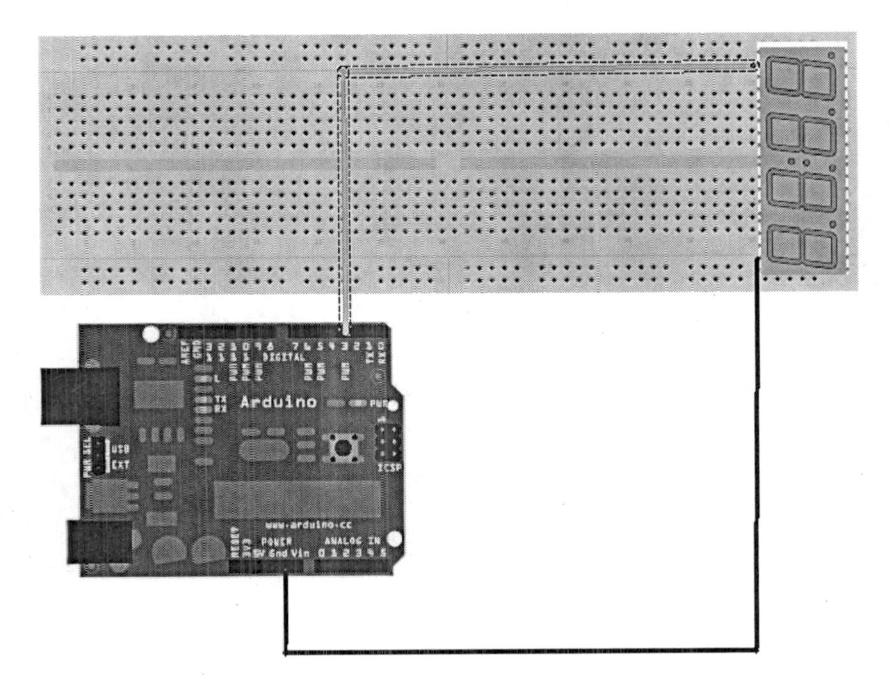

Figure 3-3. *The completed 4char test circuit.*

Step 3 Connect a jumper between Arduino digital pin 3 and the RX pin on the 4Char. Once again, put the jumper through the RX hole in the display, and anchor it to the breadboard.

Write the Code

The following sample sketch includes tips as to what your 4Char display can do. You can find it on EMWA GitHub repository | chapter-3 | 4Char (*https://github.com/ejgertz/EMWA/blob/master/chapter-3/4Char*).

Load it onto Arduino.

```
/*
  4Char Test
  Sketch in Programming to test a scrolling 4Char display.
  The traditional 'first code' is to display "Hello World", but
  the 4char can't display the letter "w". So we improvise...
  This example code is based on example code that is in the public domain.
*/

#include <SoftwareSerial.h>

#define SerialIn 2
```

```
#define SerialOut 3

#define Xdelay 600

String txtmessage = "HELLo ThErE";

byte thou=0;
byte hund=0;
byte tens=0;
byte ones=0;

SoftwareSerial mySerialPort(SerialIn, SerialOut);

void setup()
{
  pinMode(SerialOut, OUTPUT);
  pinMode(SerialIn, INPUT);


  // open communications with the 4char display
  mySerialPort.begin(9600);

  // the 'v' character resets the display
  mySerialPort.print("v");

}

void loop()
{
  // light up all segments as a test
  mySerialPort.print("----");
  delay(Xdelay);
  mySerialPort.print("8888");
  delay(Xdelay);
  mySerialPort.print("----");
  delay(Xdelay);
  mySerialPort.print("xxxx");
  delay(Xdelay);

  // scroll from 1 to 0 the simple but tedious way
  mySerialPort.print("xxxx");
  delay(Xdelay);
  mySerialPort.print("xxx1");
  delay(Xdelay);
  mySerialPort.print("xx12");
  delay(Xdelay);
  mySerialPort.print("x123");
  delay(Xdelay);
  mySerialPort.print("1234");
  delay(Xdelay);
  mySerialPort.print("2345");
```

```
delay(Xdelay);
mySerialPort.print("3456");
delay(Xdelay);
mySerialPort.print("4567");
delay(Xdelay);
mySerialPort.print("5678");
delay(Xdelay);
mySerialPort.print("6789");
delay(Xdelay);
mySerialPort.print("7890");
delay(Xdelay);
mySerialPort.print("8900");
delay(Xdelay);
mySerialPort.print("9000");
delay(Xdelay);
mySerialPort.print("0000");
delay(Xdelay);

// Count from -1009 to 2000

for(int i = -1009; i<2000; i++)
{
   if((i<-999) || (i>9999))
    {
    mySerialPort.print("ERRx");
    return;
    }
   char fourChars[5];
   sprintf(fourChars, "%04d", i);

   mySerialPort.print("v");
   mySerialPort.print(fourChars);

   //add a delay if the numbers go by too fast
   //delay(Xdelay);

}

delay(Xdelay);
mySerialPort.print("xxxx");
delay(Xdelay);

// Scroll a txtmessage a more complicated way
// First add the appropriate buffer
txtmessage = "xxxx"+txtmessage+"xxxxx";
// then convert from String object to char array,
// which is the only thing SoftwareSerial can print
char temps[txtmessage.length()];
txtmessage.toCharArray(temps,txtmessage.length());

// then scroll through the txtmessage
```

```
    for(int i = 0; i <= txtmessage.length()-5; i++)
    {
      mySerialPort.print(temps[i]);
      mySerialPort.print(temps[i+1]);
      mySerialPort.print(temps[i+2]);
      mySerialPort.print(temps[i+3]);
      delay(Xdelay);
    }

    delay(Xdelay);
    delay(Xdelay);
    delay(Xdelay);
}
```

Did it work? Way to go! You've successfully programmed a scrolling 4Char display.

Things to Try

1. The LED 4Char display product user guide (*http://www.sparkfun.com/datasheets/Components/LED/7-Segment/YSD-439AB4B-35.pdf*) from SparkFun describes how to control each individual segment of the display, as well as the colon, apostrophe, and decimal points. Can you use this knowledge to create a countdown timer that counts from 59:59 to 00:00?
2. Suppose you really needed to use this display to show the letters k, m, v, w, or x. For example, if you absolutely had to display the words *my wax Vostok* (though we can't think of why you would), how would you do it? Which segments would you light up?

4/Detecting Electromagnetic Interference (and making bad music)

The gadget we're going to build next can detect electromagnetic interference (EMI), and give you a rough idea of the intensity of the EMI signal.

EMI is a form of electromagnetic radiation: a combination of electric and magnetic waves traveling outward from anywhere that an electrical power signal is changing or being turned on and off rapidly.

Sometimes, an electrical device that has the potential to give off EMI is very carefully shielded to prevent the interference from escaping; however, a great many devices that emit EMI are shielded not at all or very lightly.

Since EMI is a type of radio signal, this gadget is essentially a type of radio. We won't be listening for any particular station or program, however. We're listening for electromagnetic energy being emitted from various electronic devices in the local environment, and converting that into an output that our human senses can detect.

Where this gadget excels is spotting "phantom" or "vampire" energy loads. More correctly called standby power, this is the amount of electricity that constantly flows through some electronic devices, even when they're supposedly switched off or in standby mode. Devices use standby power on features such as digital clocks, remote control reception, and thermometers. Relatively weak energy efficiency regulations in the United States result in many devices drawing far more wattage than they need in standby mode.

The result? Phantom load accounts for 10% or more of the average U.S. household's home energy use. In 2009, that added $15.65 billion or more to the nation's domestic electric bill (about $125 a year per household), as well as 836 million metric tons of greenhouse gas pollution to the atmosphere.

Once you've found the energy vampires, a brief guide from NRDC's Smarter Living (*http://www.simplesteps.org/articles/co2-smackdown-step-11-de fang-energy-vampires*) website and the "Energy Savers" booklet (*http:// www.energysavers.gov/tips/*) from the U.S. Department of Energy offer tips to cut back their energy use. But the easiest way to start curbing pollution and saving money is to pull the plug.

Detecting EMI Sources in the Environment

Moving the EMI detector around our home offices revealed a fascinating variety of unsuspected energy vampires. The Toshiba laptop computer we use for development gives off a phenomenal amount of EMI. The office television, a 1998 cathode ray tube model from Sony, gave off even more. This makes sense, because the TV is essentially built to give off EMI. (That lightsaber-like hum you hear from very old televisions? EMI artifacts.)

Strangely enough, the WiFi router emitted very little EMI, at least in the range that this detector can spot.

The most surprising phantom load that we found with the EMI detector came from the office stereo system, a component bookshelf model. The 4Char lit up and the speaker squealed from several feet away.

It turns out that the stereo system uses nearly as much electricity turned off as our netbook does turned on.

Demystifying Radiation

The term "radiation" often scares people when it comes to environmental monitoring, because it's used for several similar but not identical phenomena.

Radiation simply refers to something that radiates outward from a source. In the case of uranium and plutonium, the stuff being radiated outward is subatomic particles, which have been proven very dangerous to living tissue. In the case of EMI, the stuff being radiated is electromagnetic waves.

We've been living with human-created EMI radiation for 100 years, and only in the case of very high energies has it been shown to cause any harm.

Make the Gadget

This gadget is one of the simplest environmental sensors you can make. All it does is connect an antenna to one of Arduino's analog ports and output the results as numbers and sounds.

There are some things about this device and how it works that make more sense once you've actually created and used it. So we've included that information after the build.

Parts

1. Arduino
2. 8-ohm speaker
3. 4Char display
4. 1-megaOhm resistor
5. 3–5 feet of solid core wire
6. Battery pack
7. Red and black jumpers

The 8-Ohm Speaker

We're going to be using a standard 8-ohm speaker as the output. This is probably the most common kind of hobby audio output, one of the most basic (and oldest) electronic devices.

Simply put, a speaker is an electromagnet connected to a membrane. Variations in an electric current cause the electromagnet to turn on and off. This moves the membrane back and forth, which moves air molecules back and forth, which causes what we call sound.

If the movement is done rapidly enough, and if the voltage signal can be precisely controlled, the speaker will emit sound we can recognize—like music, or a person talking.

Before we begin to build the gadget, let's be sure the 8-ohm speaker works:

Step 1 Plug your 8-ohm speaker into Arduino, as shown in the breadboard view (Figure 4-1): the red wire into digital pin 8 and the black wire into GND.

Step 2 Find the sketch "ToneMelody" in the IDE at File | Examples | Digital | ToneMelody. Load it onto Arduino and run it.

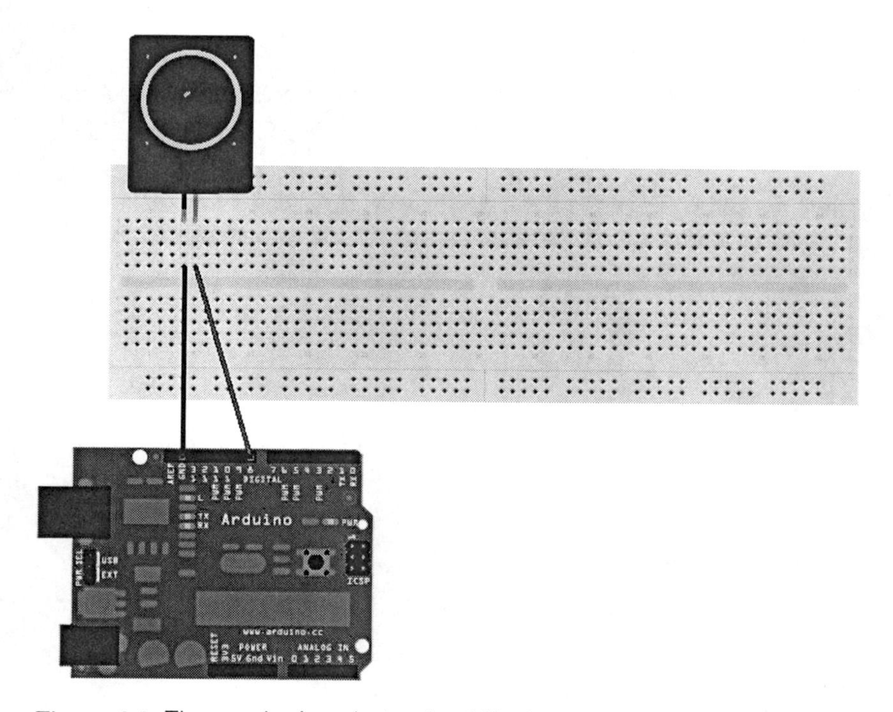

Figure 4-1. *The speaker's red wire should be inserted in digital pin 8 and the black wire in GND.*

In the sketch, the header file *pitches.h* tells the Arduino which audio frequencies correspond to which musical notes. The sketch then includes the data for a little eight-note melody, in two arrays—one containing the notes, and the other containing the duration of those notes. The sketch then plays these tones through Arduino pin 8.

Did you hear a pleasing little melody when you ran the sketch? Then it worked. If not, check your connections, and try again. Make sure that the speaker is plugged into Arduino pin 8 (and not any other pin) and GND.

Construct the EMI Monitor

Step 1 Cut a three-foot long piece of solid core wire (Figure 4-2).

Step 2 Strip about 1.5 inches from one end of the core wire.

Step 3 Wrap one end of the 1MOhm resistor around the stripped end of the core wire.

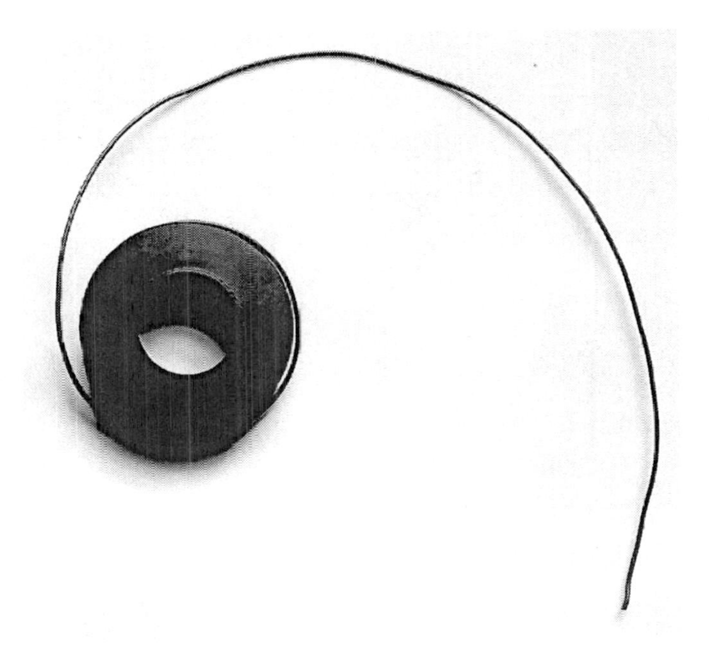

Figure 4-2. *Solid core wire, used as the "antenna" for the EMI detector.*

Step 4 Insert the wire into analog port A5 on Arduino, and the end of the resistor into GND.

Step 5 Connect one lead of the 8-ohm speaker to digital port 9 on Arduino, and the other lead into GND, as shown in the breadboard view (Figure 4-3).

Step 6 Connect the lead of the 4Char to digital pin 7 on Arduino and the other lead into GND.

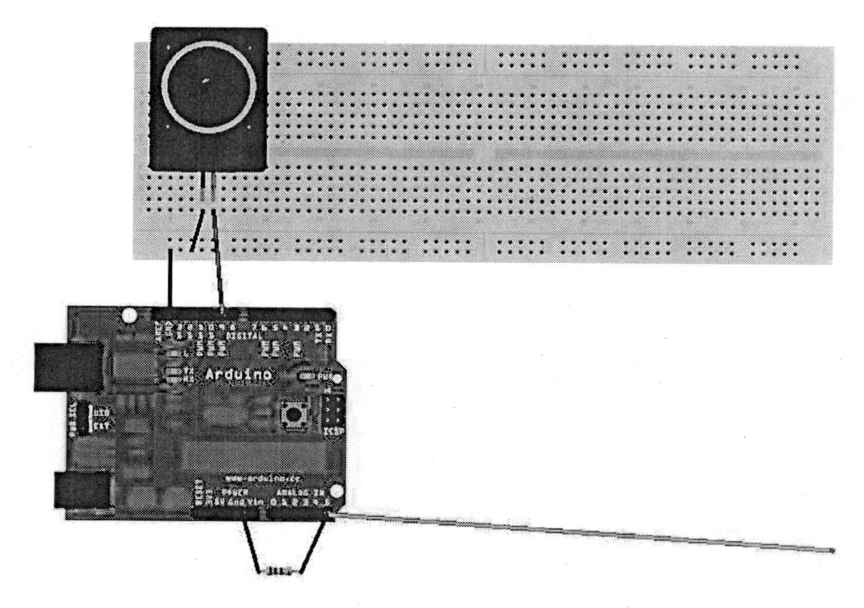

Figure 4-3. *EMI detector: one lead of 8-ohm speaker connected to GND pin on Arduino, the other to digital port 8.*

Write the Code

Load the following sketch onto Arduino, and you've got yourself an EMI detector:

```
// Arduino Electromagnetic interference detector
// Code modified by Patrick Di Justo, based on
// Aaron ALAI EMF Detector April 22nd 2009 VERSION 1.0
// aaronalai1@gmail.com
//
// This outputs sound and numeric data to the 4char

#include <SoftwareSerial.h>
#define SerialIn 2
#define SerialOut 7

#define wDelay 900

int inPin = 5;
int val = 0;

SoftwareSerial mySerialPort(SerialIn, SerialOut);

void setup()
```

```
{
  pinMode(SerialOut, OUTPUT);
  pinMode(SerialIn, INPUT);

  mySerialPort.begin(19200);
  mySerialPort.print("vv");

  mySerialPort.print("xxxx");
  delay(wDelay);
  mySerialPort.print("----");
  delay(wDelay);
  mySerialPort.print("8888");
  delay(wDelay);
  mySerialPort.print("xxxx");
  delay(wDelay);

  Serial.begin(9600);
}

void loop()
{
 val = analogRead(inPin);
 Serial.println(val);
 dispData(val);
 val = map(val, 1, 100, 1, 2048);
 tone(9,val,10);
}

void dispData(int i)
{
  if ((i<-999) || (i>9999))
  {
   mySerialPort.print("ERRx");
   return;
  }
  char fourChars[5];
  sprintf(fourChars, "%04d", i);

  mySerialPort.print("v");
  mySerialPort.print(fourChars);
}
```

Run the Sketch

Once you have uploaded your sketch to Arduino, and Arduino restarts, you'll probably hear a cacophony of sound from the speaker, and the numbers on the 4Char display will be so random as to seem meaningless. That's perfectly all right; they are meaningless.

Because Arduino is connected by a USB cable to your computer, it is receiving a flood of electromagnetic interference from the computer. Even worse, that EMI is being pumped into Arduino via the USB cable.

To make this detector really work, we've got to go mobile.

Powering the Gadget in Mobile Mode

When not operating off the USB cable, Arduino needs a power supply of 7 to 12 volts to work. Anything higher than 12 volts might damage the circuitry; anything lower than 6 volts won't start Arduino at all.

A fresh 9-volt battery should be enough to get this gadget running.

Step 1 Carefully unplug the Arduino from the USB cable.

Step 2 Snap the power connector to the top of the 9-volt battery, and plug the round end into the power input on Arduino.

Your Arduino should start up normally: the LEDs mounted on the Arduino board should flash, and within a few seconds the EMI code should be up and running.

What Are We Measuring with This Gadget?

Now that you've successfully built the mobile EMI detector, we can explain what it is measuring.

Arduino's analog input normally takes a reading of the electrical energy coming into the analog port. But because we have connected an antenna to that port, the antenna is absorbing electrical voltage from the radio signals given off by electronic equipment, and directing it into the analog port.

Arduino analog's port can take voltage from zero to a maximum of 5 volts, and it measures this voltage in 1024 discrete slices (making each slice worth 0.0048828 volts). For example, when Arduino tells us that the reading from the analog port is, say, 250, it is telling us that the antenna wire is picking up 1.2207 volts of EMI energy (250×0.0048828).

The Truth Is Out There

Do a little online research into EMI detectors (sometimes called EMF detectors), and you'll quickly discover people who claim to use electromagnetic disturbances to "detect" paranormal activity such as ghosts, poltergeists, and spirits.

Sorry to go all Scully on you, but we've yet to encounter compelling evidence that ghosts exist—or, if they do, that they obey the rules of electromagnetic propagation.

Anyone who wishes to use this EMI detector to go ghost hunting probably can't be stopped. But please: don't contact us for advice or argument, and don't say you read it here first.

The raw number from the analog port is then sent to the speaker, where it is converted to a tone, and to the 4Char, where it is displayed as a value.

The rapidly changing nature of the EMI voltage picked up by the antenna gives rise to rapidly changing tones, ranging from a low electronic growl to a high pitched electric squeal.

The voltage induced in the antenna wire is very dependent on the length of the wire. It seems pretty obvious: a longer wire can collect more voltage than a shorter one. If you want consistent results from your EMI detector, make certain that you reuse the same wire every time, or at least a wire of the same length.

Things to Try

1. What happens if you make the wire shorter? If you make the wire longer? If you coil the wire into a circle? Does the detector become more sensitive? Less sensitive?
2. What are some other ways to display the data you obtain? A row of LED lights, as in the noise sensor?

5/Project: Water Conductivity/ Numerical Output

The conductivity meter is probably the simplest environmental meter in this book. Its workings rely upon the fact that pure water does not actually carry an electric charge very well. So what we're really doing with this device is assessing the concentration of conductive particles that are floating in the (mostly nonconductive) water.

What Is Conductivity, and Why Do I Care?

Water is very seldom just the sum of its basic chemical formula: two atoms of hydrogen and one of oxygen. Typically, water is a mixture that also includes other substances that have dissolved into it, including minerals, metals, and salts. In chemistry, water is the *solvent*, the other substances the *solutes*, and combined they make a *solution*.

Solutes create ions: atoms that carry an electric charge. These ions are what actually move electricity through water.

That's why measuring conductivity is a good way to learn how pure (really, how impure) a water sample may be: the more stuff that's dissolved in the watery solution, the faster electricity will move through it.

Make the Gadget

To make this device, we'll start by constructing the probe that you'll dip into water samples to get a conductivity reading—which we created by repurposing a binding post (Radio Shack catalog #274-718 (*http://www.radioshack.com/product/index.jsp?productId=2102838*)) typically used in home audio/video and ham radio electronics. It may seem like a complicated build on first glance, because there are several steps involved. But it's mostly a straightforward matter of connecting wires to the binding post—as you can see in the photo of the finished probe.

Water and Electricity Don't Mix...Right?

All our lives we've been told to keep water and electricity far apart for our safety. Lifeguards order swimmers out of the water when lightning threatens. Electricians wire bathroom power outlets especially to prevent water-based electric shocks. We're told not to drive through flooded streets if there are downed power lines in the water.

So how can we say that water is not very conductive?

Well, it's all relative. Compared to air, water is highly conductive. One reason lightning bolts are so powerful is that they need enormous energy to overcome the electrical resistance of the air. However, when compared to metal, which conducts electricity extremely well, water is so resistant that it is practically an insulator.

Parts

1. Arduino
2. Breadboard
3. 10K resistor
4. 8-ohm speaker
5. 4Char display
6. Chassis-mount dual female binding post (RadioShack catalog #274-718)
7. Jumper wires in various colors
8. Several feet of insulated solid core wire
9. Small piece of aluminum foil
10. 4-inch adjustable wrench
11. Distilled water (can be purchased at some pharmacies or health food stores)

Construct the Probe

Check your work by comparing it to the photo of a completed probe (Figure 5-1).

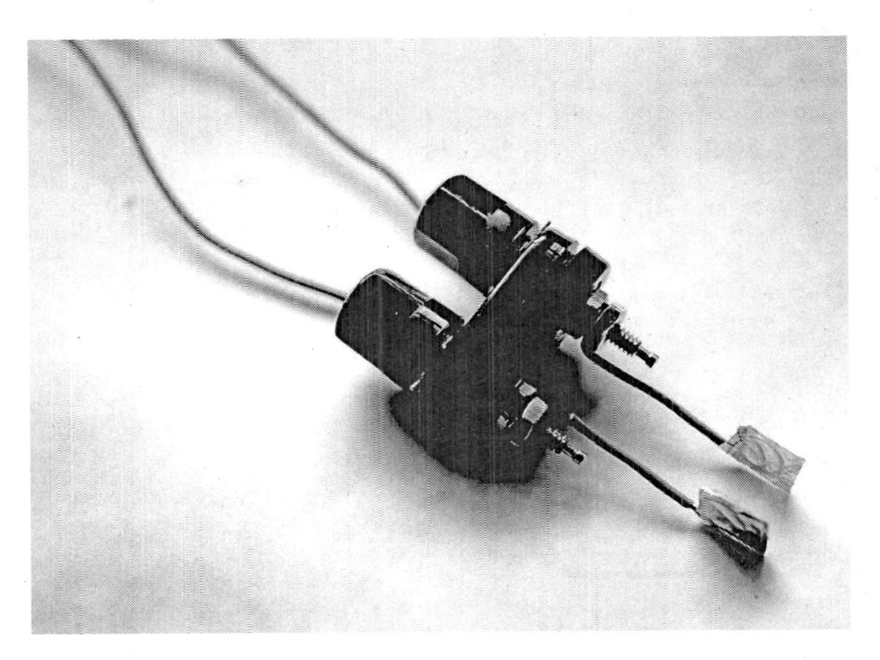

Figure 5-1. *The probe!*

Step 1 Cut two three-foot lengths of solid core wire and two four-inch lengths of solid core wire.

Step 2 Strip about an inch of insulation off both ends of each long wire. Strip one inch of insulation off one end of each short wire, and one and a half inches off the other end.

Step 3 Unscrew the binding post's black and red knobs partway (they won't come all the way off), so that you can see the hole drilled through each of the two posts.

Step 4 Insert a stripped end of one long length of solid core wire through one of the holes, and then screw that post's knob back down until it is holding the wire securely in place. Clip the tip of the wire flush with the knob, so that you won't accidentally stab yourself with it later when using the probe. Repeat with the other post and the second long length of wire.

Step 5 Using the small adjustable wrench, unscrew the outermost nuts on the opposite, bare metal ends of each post, until they are about halfway up the post (but still engaged with the threads).

Step 6 Wrap the one-inch stripped end of one short length of wire around one of the posts, in between the two nuts. Then screw the loosened nut back down until the wire is held tightly between the two nuts. Repeat for the other short length of wire and the other post.

Step 7 Bend these wires so that they extend straight past the ends of the posts.

Step 8 Shape the ends of these wires into small spirals, and then neatly fold a small piece of aluminum foil—just a few inches—over and around each spiral to cover. Position these two "paddles" so that their flat surfaces are parallel to each other and about 1 centimeter apart.

Breadboard the Circuit

Check your work by comparing it to the breadboard view of the completed circuit (Figure 5-2).

Step 1 Connect one jumper from the probe to Arduino analog pin A5. Connect the other to a row in the breadboard.

Step 2 Add a 10K resistor (color code brown-black-orange) to the breadboard, in the same row as the jumper from the probe. Have the resistor bridge the gap between both sides of the breadboard.

Step 3 Connect a ground jumper from the GND pin on Arduino to the breadboard, in the same row as the 10K resistor.

Step 4 Connect a data jumper from the breadboard to Arduino Analog pin A0. Note carefully that the data jumper starts on the same row as the resistor and GND jumper, but *before* the resistor. This is extremely important. The circuit will not work otherwise.

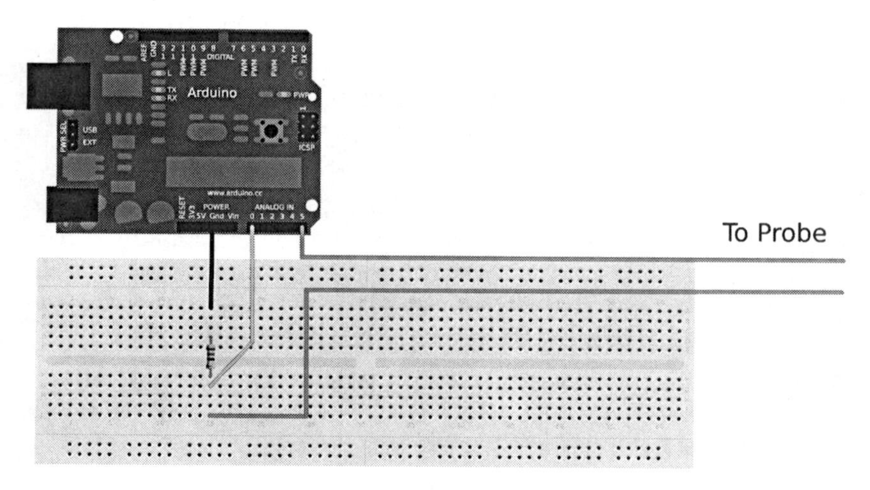

Figure 5-2. *The completed water conductivity circuit.*

Write the Code

You can find this code on the EMWA GitHub repository | chapter-5 | Water-Conductivity (*https://github.com/ejgertz/EMWA/tree/master/chapter-5*).

Load the following sketch onto Arduino.

```
/*
  Water Conductivity Monitor
  Sketch for an Arduino gadget that measures the electrical
  conductivity of water.
  This example code is based on example code that is in the public domain.
*/

const float ArduinoVoltage = 5.00; // CHANGE THIS FOR 3.3v Arduinos
const float ArduinoResolution = ArduinoVoltage / 1024;

const float resistorValue = 10000.0;
int threshold = 3;

int inputPin = A0;
int ouputPin = A5;

void setup()
{
  Serial.begin(9600);
  pinMode(ouputPin, OUTPUT);
  pinMode(inputPin, INPUT);
}

void loop()
{
  int analogValue=0;
  int oldAnalogValue=1000;
  float returnVoltage=0.0;
  float resistance=0.0;
  double Siemens;
  float TDS=0.0;

  while(((oldAnalogValue-analogValue)>threshold) || (oldAnalogValue<50))
  {
    oldAnalogValue = analogValue;
    digitalWrite( ouputPin, HIGH );
    delay(10); // allow ringing to stop
    analogValue = analogRead( inputPin );
    digitalWrite( ouputPin, LOW );
  }

  Serial.print("Return voltage = ");
  returnVoltage = analogValue *ArduinoResolution;
```

```
    Serial.print(returnVoltage);
    Serial.println(" volts");

    Serial.print("That works out to a resistance of ");
    resistance = ((5.00 * resistorValue) / returnVoltage) - resistorValue;
    Serial.print(resistance);
    Serial.println(" Ohms.");

    Serial.print("Which works out to a conductivity of ");
    Siemens = 1.0/(resistance/1000000);
    Serial.print(Siemens);
    Serial.println(" microSiemens.");
    Serial.print("Total Dissolved Solids are on the order of ");
    TDS = 500 * (Siemens/1000);
    Serial.print(TDS);
    Serial.println(" PPM.");
    if (returnVoltage>4.9) Serial.println("Are you sure this isn't metal?");

    delay(5000);
}
```

How to Take a Reading

To use this device, first open a "serial monitor" window in the Arduino IDE. This is where your water conductivity readings will appear.

Pour some of the distilled water into a glass, and set it down a couple feet away from your computer, Arduino, and circuit, such that they wouldn't be damaged if the glass were to accidentally tip over.

Dip the probe's two "paddles" into the water, and watch the serial monitor window for the readings to appear. Since distilled water is nearly free of solutes, the conductivity should be quite low.

To take further readings on this or other water samples, simply press the reset button on Arduino.

Things to Try

1. Build a conductivity meter without Arduino. This small gadget will give you a very broadly accurate (i.e., not very accurate at all) idea of the conductivity of a sample of water: Simply connect an LED to a 1-megaohm resistor to a 9-volt battery. This will give you a binary sense of the water's electrical conductivity: if the LED lights up, the water is conductive. if it doesn't, the water is not very conductive (we told you it wasn't very accurate).

2. Improve the accuracy of this device by calibrating the gadget and using different resistors in the circuit. Take a known quantity of distilled water (say, 1000 milliliters) and add a known quantity of salt (say 1 gram) to it. That will give you a calibration solution of 1000 parts per million. By using different resistors until the LED just comes on, you can make a conductivity meter that checks for water with varying levels of dissolved solids.
3. Try making different calibration solutions and using different resistors to find when the LED comes on. You'll end up with a chart of different conductivities, and highly accurate ways to test for them.

6/New Component: Ethernet Shield

Arduino does many interesting things on its own, but it can't do everything. Fortunately, thanks to the open design of Arduino, building add-on boards to extend what Arduino can do is pretty easy.

These boards are called "shields," because they usually fit over the top of Arduino like a protecting shield.

Shields contain their own circuitry, components, and sometimes processing chips that can augment Arduino's capabilities. There are shields that enable Arduino to read GPS signals, drive robot motors, take pictures, connect to Bluetooth devices, control electroluminescent wire...as of this writing, there are more than 300 shields. Keep up to date on shields at *http://shieldlist.org*.

In our opinion, the Ethernet shield (Figure 6-1 and Figure 6-2) is the most important Arduino shield, and the first one you should make or buy. Hooking your Arduino to an Ethernet connection can put it on the Internet, and once that's done your options become nearly endless. Want to use your Arduino to set up a small web server? No problem. Place data online for other people to see? Simple. Send tweets via Twitter? Easy breezy!

Notice how the Ethernet shield is shaped very much like Arduino? It's been built that way to be compatible: the pins in an Ethernet shield reach downward into Arduino's sockets and make a firm connection.

The Ethernet shield also has sockets with the same layout and the same functions as Arduino itself. This means that you can plug the shield into Arduino and then forget it. Each gadget in this book can be plugged into an Arduino+Ethernet shield combo as easily as into Arduino itself. In fact, usually we keep our Ethernet shield plugged into Arduino at all times, and develop our gadgets that way.

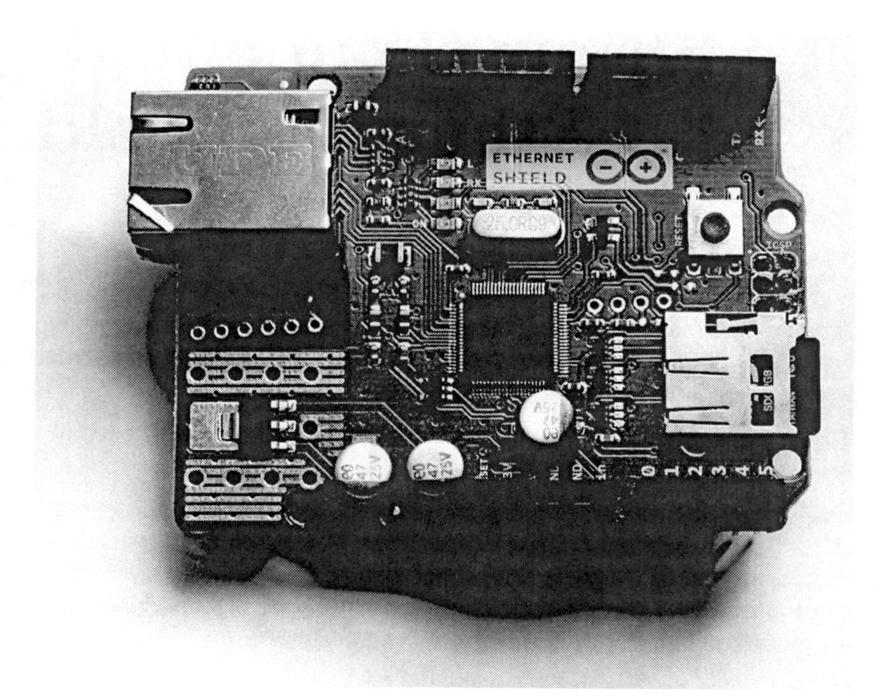

Figure 6-1. *Front of the Ethernet shield.*

Using the Ethernet Shield

Let's take a look at the Ethernet shield's special features.

The Ethernet Port

The Ethernet shield we use comes from SparkFun (sku DEV-09026 (*http://www.sparkfun.com/products/9026*)). It requires a standard RJ-45 cable (the cable that looks like a fat landline phone cord). One end of the RJ-45 cable plugs into the shield, while the other end plugs into your Internet router or possibly your cable modem. It will most likely be plugged into your router.

The MAC Address

At the bottom of the shield, or in the packaging, you should see a sticker printed with a cryptic sequence of letters and numbers, something like "90-A2-DA..." and so on. This is the shield's Media Access Control (MAC) address.

Figure 6-2. *Back of the Ethernet shield.*

A MAC address is a completely unique identifier for anything that is going to be connected to the Internet. Everything you own that connects to the Internet, from your computer to your digital music player to your mobile phone, has a unique MAC address.

For the devices that you don't program, you normally don't need to be concerned with the MAC address. But in this book, you'll be programming your Ethernet shield, so knowing the MAC address is vital.

So write down the MAC address of your Ethernet shield somewhere safe, and DO NOT REMOVE THE STICKER. Since all Ethernet shields look alike, the easiest way you'll be able to tell one from another is by reading the MAC address sticker. An Ethernet shield without a known MAC address is an annoyance you don't want to deal with.

The IP Address

You'll also need to know the Internet Protocol (IP) address for your router. This is the number that gets assigned to every device on a local network. For most routers sold in North America, the default IP address is 192.168.1.1.

(Apple routers often default to an IP address of 10.0.1.1, because that's how Apple rolls.)

The good thing about this is that if you don't know what your router address is, and you are not using an Apple product, 192.168.1.1 is probably the correct IP address.

Find your router's IP address: Windows

1. Click on Network Connections. (See Figure 6-3.)
2. Right-click a network connection.
3. Click Status, and then click the Support tab.
4. The "Default gateway" is the router's IP address, which in this case is indeed 192.168.1.1.
5. Also, take note of the Subnet Mask (255.255.255.0). You might need it later on.

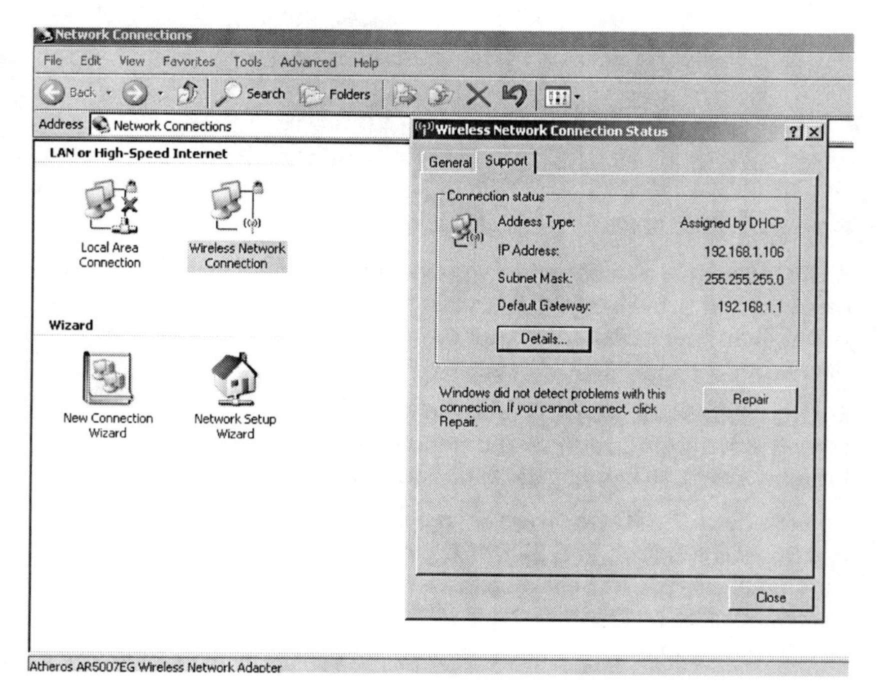

Figure 6-3. *The network connections panel in Windows.*

Find your router's IP Address: Mac

1. Open System Preferences.
2. Under the options for Internet & Wireless, select Network.
3. On the Network panel, be sure that your current Internet connection method (likely either AirPort or Ethernet) is selected in the left menu, and that your current location is selected on the Location pull-down menu.
4. Still on the Network panel, click the Advanced button, located toward the lower right.
5. A list of networks will drop down into view, with a row of options along the top. On this row, find and click TCP/IP.
6. You will see your Router address, which is indeed 192.161.1.1, as shown in Figure 6-4.
7. Also, take note of the Subnet Mask (255.255.255.0). You might need it later on. (You can ignore the IPv4 address.)

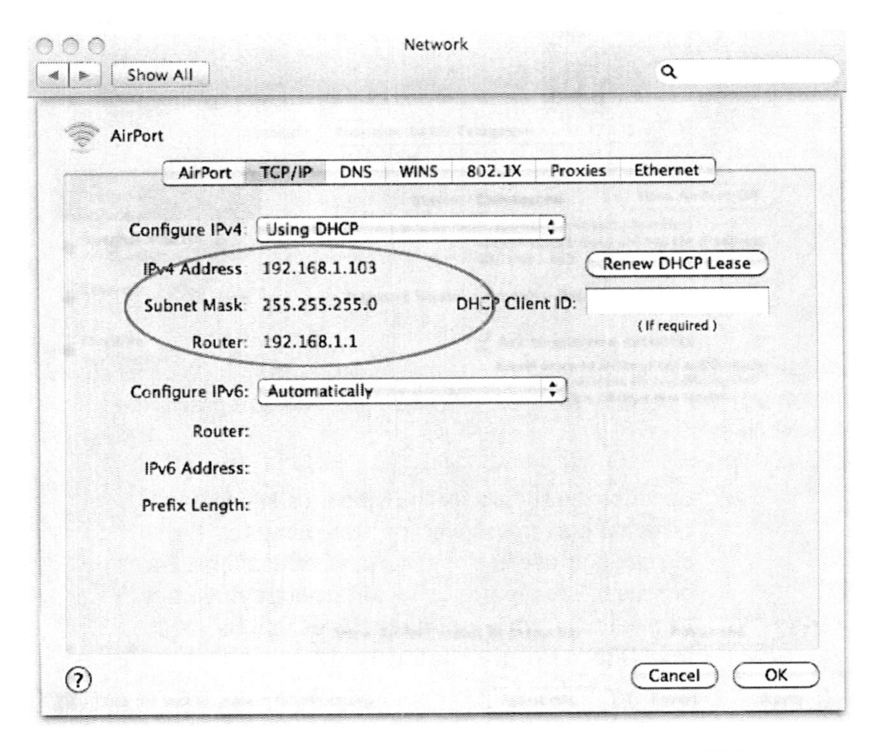

Figure 6-4. *The Advanced Network preferences panel in Mac OS X.*

The SD Card Slot

The SD card slot is a wonderful bonus of the Ethernet shield. It allows you to write and read (store and retrieve) data to and from a microSD card. We've tested ours with a 4 GB microSD card, and it works like a charm.

You may want to buy a microSD card that comes with a full-size SD card adapter (Figure 6-5), or buy an adapter separately, because this will enable you to swap the card back and forth between Arduino and conventionally sized card readers or slots.

Figure 6-5. *The MicroSD card can be moved between your Ethernet shield and an adapter sized for the typical SD card slot on a computer.*

Because the signals for the SD card slot and the Ethernet port travel over the same circuitry, Arduino can only use one of these devices at a time. For practical purposes, this shouldn't matter much, but be aware of it while writing sketches that use both the SD card slot and the Ethernet port.

Testing the Ethernet Shield

The Ethernet shield is an essential part of the rest of the gadgets in this book. So let's test it to be sure it works before building anything with it.

Parts

1. Arduino
2. Ethernet shield

Assembly

Step 1 Attach the Ethernet shield to Arduino.

Step 2 Plug the shield into your Internet router.

Step 3 Find the sketch "Web Server" in the Arduino IDE, under File | Examples | Ethernet | WebServer, or on EMWA GitHub repository | chapter-6 | WebServer (*https://github.com/ejgertz/EMWA/blob/master/chapter-6/WebServer*). Load it onto Arduino.

Be certain to insert the MAC address of *your* Ethernet shield and the IP address you wish to give to *your* Arduino into the following sketch. A good rule of thumb is to add 20 to the IP address of your router, which should ensure that your Arduino doesn't have the same address as other devices on your local network (as long as you have no more than 20 devices on the network.) That is, if your router is at IP address 192.168.1.1, make your Arduino's IP address 192.168.1.21.

Step 4 When you've uploaded the sketch to Arduino, and the sketch is running, go to your computer, open up your Internet browser (Chrome, Safari, Firefox, Internet Explorer, or whichever), and enter your Arduino's IP address into the address bar.

You should see a very simple little web page, resembling what the Web looked like in 1994. Congratulations! Your Arduino is now a web server on the Internet.

Testing the SD Card Slot

Now we want to test the SD card slot, since it can be used in all the subsequent gadgets in this book.

Parts

1. Arduino
2. Ethernet shield
3. microSD card

Assembly

Step 1 Insert the microSD card into the holder.

Step 2 Load the sketch "SD card read/write," which you can find in the IDE under File | Examples |SD |ReadWrite or on EMWA GitHub repository | chapter-6 | SDCardReadWrite (*https://github.com/ejgertz/EMWA/blob/master/chapter-6/SDCardReadWrite*).

Step 3 If the Arduino successfully reads "testing 1, 2, 3" from the SD card and displays it on the serial output, the SD card slot works.

If instead, the serial output displays the words "error opening test.txt," there is an error somewhere and you'll need to do some troubleshooting.

Things to Try

1. Can you write a sketch that displays meaningful data on a web page?
2. Can you write a sketch that displays data on a web page *and* saves it to the SD card?

7/Project: Humidity, Temperature & Dew Point/4Char Display

Weather may be the most fundamental experience we have of the natural world. Every day we make decisions based on present and near-future environmental conditions: Do I need a coat? Should I bring an umbrella? Can I bicycle to work, or do I need to catch the train?

You Don't Have to Be a Weatherman to Measure the Weather

We put a lot of effort into controlling how the weather affects us. We build structures that shield us from rain, snow, wind, and sun. We create all sorts of specialty textiles, and use them in complex garments to protect ourselves from the elements when we go outdoors. And we use lots of electricity to power systems that keep us comfortably warm or cool, as well as devices that put moisture into the air or suck it dry.

Weather is so important to our personal and economic well-being that a government agency is devoted to doing nothing but predicting and tracking the weather and keeping the public informed. One of the most popular channels on cable TV devotes almost every hour of every day to reporting nothing but weather. Dozens of computer, smartphone, and tablet apps exist to do one thing: hook us into the latest info about the weather.

In other words, it's high time makers take on some independent monitoring of the weather. With Arduino, we can measure the basics: temperature, humidity, and dew point. Further, we can collect these measurements over time to create a weather record by saving the data to an SD card, using the Ethernet shield's SD card slot.

What's Dew Point?

The dew point, simply put, is the temperature at which moisture condenses to form dew. A low dew point temperature indicates drier conditions, while a higher dew point temperature means wetter conditions.

To find the dew point, our device takes current temperature and humidity readings, and performs a calculation to determine how saturated the air is with water.

Dew point is a particularly important weather factor in transportation, because in low temperatures it can help predict whether ice will form on airplane exteriors, roads, or train tracks.

Getting Usable Measurements

There are a lot of variables involved in getting accurate temperature readings. Are you measuring the temperature indoors or outdoors? In a closed stuffy room, or in one with lots of ventilation? Over a black asphalt driveway or over relatively cool white concrete, in full sunlight or in the shade?

These widely varied conditions can result in wildly different readings, even if they're taken at the exact same moment.

Weather professionals have developed a method to ensure they're all measuring temperature the same way across multiple outdoor locations. They place a thermometer in a Cotton Region Shelter (CRS), a white-painted pinewood box with a solid top, slatted sides, and a slotted bottom, and place it five to six feet off the ground.

The CRS protects the thermometer from sun, rain, snow, and more—from everything except air temperature.

You don't need a CRS in order to use this gadget (though building one would be an awfully cool project—here are some sample specs (*http://www.weather.gov/pa/faq.php#q17*) from the National Weather Service). But we do recommend that you find a way to create consistent conditions when you use it.

Indoors, this should be fairly easy. Find a place to keep the gadget that emulates the advantages of a CRS: out of direct sunlight, away from heating or cooling vents, several feet off the ground.

Outdoors, however, you're going to need something more like a CRS, something that is waterproof but allows for circulation of air around the sensor.

Weather and Climate: What's the Difference?

The next time the ground somewhere in the country is covered in drifts of snow, and temperatures stay frigid for several days on end, it probably won't be long before you hear a TV pundit say something like, "Get a load of this cold! So much for global warming."

This confused pundit is giving voice to a common misunderstanding: the difference between weather and climate.

Weather is the blanket term for the immediate atmospheric conditions in the environment right around us right now, and in the coming hours and days—cloud cover, wind, temperature, precipitation, and more.

Climate, on the other hand, is a concept that's much broader in both time and space. Climate is the aggregate over several years of a geographic area's temperature, precipitation, humidity, and more. In climate science, researchers typically look at weather patterns as well as other factors (such as ice cover in the Arctic) over a decade or longer to find trends.

When we discuss the impacts of anthropogenic (human-caused) climate change, we're talking about weather trends that have been tracked over years and decades.

The major driver of human-propelled climate change is the buildup in the atmosphere of heat-trapping greenhouse gas pollution from power plants, transportation, burning forests, and manufacturing. The sun's heat can move through these greenhouse gases to reach the Earth's surface, but they prevent some of the reflected heat from going back into space.

A little of this is a good thing: The buildup of heat-trapping gases in the atmosphere helped warm ancient Earth enough to support living organisms. But human activities since the Industrial Revolution of the mid-1700s have put much more greenhouse gas into the atmosphere than at any time in human history, increasing the Earth's surface temperature just enough to alter weather and climate conditions, in ways we're still trying to completely understand.

Citizen Climate and Weather Science

Gathering sufficient weather data to understand climate trends can be a scientist's life work. But researchers sometimes turn to the public for help gathering data, by creating "citizen science" collaborations. These can range from digitizing old weather records, to making observations outdoors, to taking environmental measurements.

One good resource for finding these collaborations is SciStarter (*http://scistarter.com/*). Search on "Climate and Weather" to find projects; there are typically dozens to choose from.

First Electronic Sensor: The DHT-22

This gadget requires a manufactured electronic sensor, since there is not much point to building a device to measure warmth, coldness, and humidity unless it returns data in standard, precisely calibrated units of measurement.

The sensor we selected is the DHT-22 (Figure 7-1), manufactured by Aosong Electronics of China. The DHT-22 uses a polymer capacitor to sense the temperature and humidity, measuring the temperature of the air between −40 and 80 degrees Centigrade (which Arduino can convert to Fahrenheit), and the relative humidity between 0 and 100%.

The necessary calibration information is stored in a tiny 8-bit computer inside the DHT-22, and each unit is tested in the factory. In other words, this sensor is ready to use right out of the box.

The DHT-22 has four pins, but only three are used. When looking at the sensor face-on, the leftmost pin is for voltage to power the sensor (anywhere from 3.3 to 6 volts; we'll use the 3.3 volt pin on Arduino); the second pin outputs data from the sensor to the Arduino; the third pin is null (not connected to anything); and the rightmost pin is GND.

Using Code Libraries

With this project, we start doing something new: using code libraries.

A library is a chunk of code that is specifically written to do a common task. Sensor code libraries give makers the means to access the functions of a sensor (as well as other electronic components) fairly confidently, because the code has been tested by the author and updated by people who use it.

Figure 7-1. *The DHT-22 sensor*

Many sensors, including the DHT-22, have their own sensor code libraries, found on repositories like GitHub (*https://github.com*). The DHT-22's library (*https://github.com/nethoncho/Arduino-DHT22*) was written and placed on GitHub by Ben Adams for hobbyists to use in building microprocessor-based gadgets that include this sensor. This saves everyone the trouble of duplicating his effort. (We've also forked a DHT-22 code library (*https://github.com/ejgertz/Arduino-DHT22*) from Ben Adams's repository, containing updates that will enable the sensor to work with the autumn 2011 Arduino 1.0 code update.)

Before you tear your hair out writing a sensor code library from scratch, search around to see if one already exists. You'll be glad you did. If you do end up writing an original code library, consider sharing it with fellow makers on a site like GitHub.

Make the Gadget

This build has several finicky steps. Look at the breadboard view (Figure 7-2) to see how it all fits together.

Parts

1. Arduino
2. Ethernet shield

3. Breadboard
4. DHT-22 sensor
5. 4Char display
6. 4.7K resistor
7. Red and black jumper wires

Breadboard the Circuit

Step 1 Power the breadboard rails.

Step 2 Connect a red jumper wire from the 3.3v pin on the Arduino to the voltage (red) rail on the breadboard, and a black jumper from the GND pin to the ground (black) rail on the breadboard.

Step 3 Connect a short red jumper from the voltage rail to the voltage (leftmost) pin of the DHT-22 temperature and humidity sensor.

Step 4 Connect a short black jumper from the GND rail to the GND (rightmost) pin of the DHT-22.

Step 5 Connect a green jumper from the DATA (second from the left) pin on the sensor to digital pin 4 on the Arduino.

Step 6 Add a 4.7K resistor between the DATA pin of the sensor and the VOLTAGE pin.

Step 7 Connect a green jumper from Arduino pin 3 to the RX (rightmost) pin of the four-character display, and a black jumper from the GND rail to the GND (leftmost) pin of the four-character display.

Write the Code

Load the following sketch onto Arduino. You can find it at EMWA GitHub repository | chapter-7 | TempHumidDewpoint (*https://github.com/ejgertz/EMWA/tree/master/chapter-7*).

```
/*
  Temperature-Humidity-Dew Point Monitor
  This sketch gathers temperature and humidity data via a DHT22 sensor,
  and also calculates dew point based on those measurements.
*/

#include <SoftwareSerial.h>
#include <stdlib.h>

#define DHT22_ERROR_VALUE -99.5
```

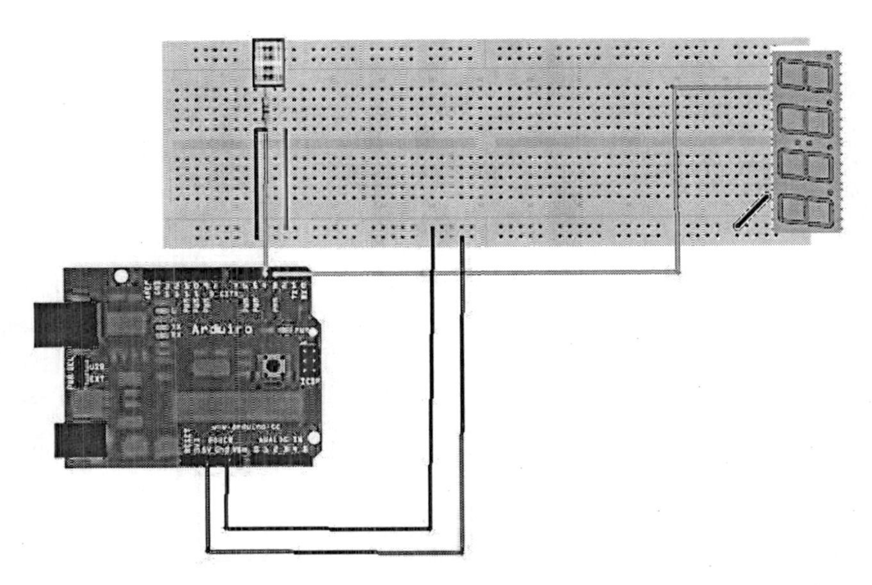

Figure 7-2. *The completed circuit for the temperature-humidity-dew point monitor.*

```
#define DHT22_PIN 4

typedef enum
{
  DHT_ERROR_NONE = 0,
  DHT_BUS_HUNG,
  DHT_ERROR_NOT_PRESENT,
  DHT_ERROR_ACK_TOO_LONG,
  DHT_ERROR_SYNC_TIMEOUT,
  DHT_ERROR_DATA_TIMEOUT,
  DHT_ERROR_CHECKSUM,
  DHT_ERROR_TOOQUICK
} DHT22_ERROR_t;

class DHT22
{
  private:
    uint8_t _bitmask;
    volatile uint8_t *_baseReg;
    unsigned long _lastReadTime;
    float _lastHumidity;
    float _lastTemperature;

  public:
    DHT22(uint8_t pin);
    DHT22_ERROR_t readData(void);
    float getHumidity();
```

```
    float getTemperatureC();
    void clockReset();
};

// Set up a DHT22 instance
DHT22 myDHT22(DHT22_PIN);

#define SerialIn 2
#define SerialOut 3
#define WDelay 900

byte thou=0;
byte hund=0;
byte tens=0;
byte ones=0;

SoftwareSerial mySerialPort(SerialIn, SerialOut);

void setup(void)
{
// start serial port
Serial.begin(9600);
Serial.println("DHT22 Library Demo");
pinMode(SerialOut, OUTPUT);
pinMode(SerialIn, INPUT);
mySerialPort.begin(9600);
mySerialPort.print("v");
mySerialPort.print("xxxx");
delay(WDelay);
mySerialPort.print("----");
delay(WDelay);
mySerialPort.print("8888");
delay(WDelay);
mySerialPort.print("xxxx");
delay(WDelay);
}

void loop(void)
{
  float tempC;
  float tempF;
  float humid;
  float dewPoint;

  DHT22_ERROR_t errorCode;

  delay(2000);
  errorCode = myDHT22.readData();
  Serial.print(errorCode);

  switch(errorCode)
```

```
{
  case DHT_ERROR_NONE:
      Serial.print("Temperature: ");
      tempC = myDHT22.getTemperatureC();

      Serial.print(tempC);
      Serial.print("C Humidity: ");

      dispData((int)tempC, 'C');

      tempF = (tempC*1.8)+32;

      delay(WDelay);
      dispData((int) tempF, 'F');
      delay(WDelay);

      humid = myDHT22.getHumidity();

      Serial.print(humid);
      Serial.println("%");

      dispData((int)humid, 'h');
      delay(WDelay);

      dewPoint = calculateDewpoint(tempC, humid);
      dispData((int) dewPoint, 'd');
      Serial.print(dewPoint);
      Serial.println("d");
      delay(WDelay);
      break;
  case DHT_ERROR_CHECKSUM:
      Serial.print("Error Cheksum");
      break;
  case DHT_BUS_HUNG:
      Serial.print("Bus Hung");
      break;
  case DHT_ERROR_NOT_PRESENT:
      Serial.print("Not Present");
      break;
  case DHT_ERROR_ACK_TOO_LONG:
      break;
  case DHT_ERROR_SYNC_TIMEOUT:
      break;
  case DHT_ERROR_DATA_TIMEOUT:
      break;
  case DHT_ERROR_TOOQUICK:
      break;
  }
}

float calculateDewpoint(float T, float RH)
```

```
{
  // approximate dew point using the formula from
  // Wikipedia's article on dew point
  float dp = 0.0;
  float gTRH = 0.0;
  float a = 17.271;
  float b = 237.7;

  gTRH = ((a*T)/(b+T))+log(RH/100);
  dp = (b*gTRH)/(a-gTRH);

  return dp;
}

void dispData(int i, char c)
{
  if (c == 'k' || c=='K' || c=='m' || c=='l' || c == 'v' ||
      c=='V' || c=='W' || c=='Z' || c=='w' || c=='z')
  {
    mySerialPort.print("bAdx");
    return;
  }

  if ((i<-999) || (i>9999))
  {
    mySerialPort.print("ERRx");
    return;
  }

  mySerialPort.print("v");

  if (i > 999) { // i between 1000 and 9999 inclusive
    mySerialPort.print(i, DEC);
  } else if (i > 99) { // i between 100 and 999, inclusive
    mySerialPort.print(i, DEC);
    mySerialPort.print(c);
  } else if (i > 9) { // i between 10 and 99 inclusive
    mySerialPort.print(i, DEC);
    mySerialPort.print("x");
    mySerialPort.print(c);
  } else if (i > 0) { // i between 1 and 9 inclusive
    mySerialPort.print("x");
    mySerialPort.print(i, DEC);
    mySerialPort.print("x");
    mySerialPort.print(c);
  } else if (i < -99) { // i between -100 and -999, inclusive
    mySerialPort.print(i, DEC);
    mySerialPort.print(c);
  } else if (i < -9) { // i between -10 and -99, inclusive
    mySerialPort.print(i, DEC);
    mySerialPort.print(c);
```

```
  } else if (i < 0) { // i between -1 and -9 inclusive
    mySerialPort.print(i, DEC);
    mySerialPort.print("x");
    mySerialPort.print(c);
  }
}
```

Things to Try

1. Write your weather data to a web page and/or an SD card. Modify the code examples found in the Arduino IDE under Examples | Ethernet | WebServer, and Examples | SD | Datalogger to work with our sketch.
2. If you create the webpage, adjust your network settings so that the whole world can see it, instead of only the computers on your local network.

8/Real-Time, Geo-Tagged Data Sharing with Pachube

Your weather data doesn't have to be lonely. Pachube is a London-based database service center through which developers (like you) can connect geographically-oriented sensor data to the Web and share it with others.

(Yes, it looks like it should be pronounced "patch-OOO-bee," but it is in fact pronounced "patch-bay." This is a geek inside joke of such depth and complexity it would take several pages to explain it. The short version is that in engineering, a *patch bay* is a place to hook lots of different electronic devices together, and that's exactly what Pachube does.)

Through the use of fairly simple application programming interfaces (APIs), Pachube can take data from your environmental monitoring device, store it online, and graph it for others to see online. If enough people put enough data into Pachube, a great deal of information can emerge.

Pachube was primarily a geek thing until March 2011, when a major earthquake and tsunami led to reactor meltdowns at Japan's Fukushima nuclear power station. Unhappy with the radiation information they were getting from their government, many Japanese makers did exactly what you're about to do—they connected their Geiger counters to the Internet, geo-tagged the readings, and shared that data online. This enabled them to create and share radiation maps of Japan quickly and independently of data provided by either the power utility or the government.

Test Project: Connecting and Uploading Data to Pachube

To get started using Pachube, you must first open up a Pachube account. We'll walk through that below, and then upload some test data to the site.

Parts

1. Arduino
2. Ethernet shield
3. Ethernet cable

Open a Pachube Account

Step 1 Go to the Pachube home page (*http://www.pachube.com*) and click on Sign Up Now.

Step 2 For now, choose the free Pachube Basic plan.

Step 3 Fill out the registration form with a username, email, password, and verification.

Step 4 When you receive the account verification email from Pachube, click on the enclosed link. It should bring you back to the Pachube site.

Step 5 Once back on Pachube, click on Create a New Link, and fill in the data for your feed.

Online Privacy

You don't have to set a location marker on Pachube. If you do, it might not always be the wisest thing to indicate your precise location to people on the Web, especially if you will be gathering data from your house. If you are under 18, check with your parents about what to do; if they don't want you revealing your home to people on the Internet, don't. If you are over 18, use your best judgment. One way to keep your data useful while maintaining your privacy is to place your location marker near, but not precisely on, your actual location.

Step 6 When you're done filling in the data, click Save.

Step 7 While still on Pachube, under My Account, click on My API Keys.

Under Your Master API Key, you're going to see a 43-character jumble of uppercase and lowercase letters, numbers, and the occasional punctuation mark. This is your API key, and it is like saying "open sesame" to the Pachube servers. When your Arduino transmits this code, Pachube will know that it belongs to you and no one else, and it will give your Arduino full access to update your feeds.

The API key is as powerful as a password, and it should not be shared with anyone you wouldn't trust with your password. Specifically, if you're asking someone for help with Pachube, *never* include your API key in demo code, email it to anyone, or post it in a forum.

With your Pachube API key in hand, as well as your Ethernet IP address and MAC address (from your Ethernet shield set-up in Chapter 6), you're now ready to test a data upload to Pachube.

Write the Code

Now you'll test your connection to Pachube.

Step 1 Carefully put your Arduino and Ethernet shield together.

Step 2 Connect an Ethernet cable from the port on the shield to a port on your router.

Step 3 Load the sketch "Pachube Sensor Client" onto Arduino. You can find it in the IDE under FILE | EXAMPLES | PACHUBE CLIENT or on EMWA GitHub repository | chapter-8 | PachubeSensorClient (*https://github.com/ejgertz/EMWA/blob/master/chapter-8/PachubeSensorClient*). As always, substitute information about *your* components and *your* Pachube account for the placeholders in the sketch:

1. Replace the MAC address in the code with your Ethernet shield's MAC address.
2. Replace the Ethernet IP address in the code with your Ethernet IP address.
3. Replace the gateway numbers in the code with your router address.
4. Replace YOUR_FEED_HERE with the number of your feed.
5. Replace YOUR_KEY_HERE with the 43-digit Pachube API key. It would probably be better to cut and paste it from the Pachube website than to type it in by hand.

Step 4 Run the sketch. If some data appears on the screen, congratulations: you've made the connection to Pachube!

If nothing happens, run through Steps 1–4 again to troubleshoot your build and code:

- Check that the Arduino and Ethernet shield are joined correctly.
- Be sure you're using the correct MAC address, Ethernet IP address, and other information about your components and your Pachube account.

- Check that you've correctly inserted the information about your components and your Pachube account into the sketch, and removed the placeholders from the sketch.

Things To Try

Share some location-specific temperature and humidity data on Pachube.

9/Project: Radiation Counter/ Sharing Data on the Internet

Let's be blunt: measuring radiation levels in the environment is a tricky business, usually best left to the professionals. It's easy to come up with data that will scare you for no good reason, and it's a challenge to compare your data in useful ways to data collected by others.

What's more, few of us will ever face the risk of being exposed to excessive amounts of radiation.

In this chapter, we are talking about atomic radiation, not the electromagnetic radiation covered earlier in this book. These are different phenomena with similar names. For a refresher, see the side bar on Demystifying Radiation in Chapter 4.

So, why are we going to teach you how to build your own Geiger counter with Arduino, and how to share your readings online with people around the world?

First, because it's a fun and challenging thing to do. Second, because recent history shows that the professionals don't always plan for everything, leaving gaps that makers can help fill. As we mentioned in the previous chapter, that's what happened in March 2011, after northeastern Japan was hit by a 9.0 earthquake, followed minutes later by a 49-foot-high tsunami.

The twin disasters knocked out grid and backup electricity, respectively, to the Fukushima One (Fukushima Dai'ichi) nuclear power station in Futuba, located about 150 miles north of Tokyo. This shut down crucial cooling systems for the reactors, as well as pools in the reactor buildings holding super-hot spent fuel rods. The devastating wave also destroyed a network of radiation monitors around the plant.

Fukushima One's three working reactor cores, as well as their spent fuel pools, overheated far past their safety points, leading to a complex nuclear crisis (one beyond the scope of this book to explain) and several leaks of dangerous radioactive materials into the atmosphere. Lacking nearby radiation monitors, officials had trouble assessing the releases, leaving people in Japan and around the world understandably angry and frightened for their health and the environment.

On both sides of the Pacific, people who happened to have Geiger counters began monitoring radiation levels wherever they were, comparing them to government data, and sharing this information online with systems like Pachube.

As we write, these networks are still going and growing. Hackers who collaborated over the Fukushima disaster (Emily covered this in an April 2011 article for *OnEarth Magazine*, "Got iGeigie? Radiation Monitoring Meets Grassroots Mapping" (*http://www.onearth.org/blog/citizen-radiation-monitoring-meets-grassroots-mapping*)) have created the Safecast (*http://blog.safecast.org/*) citizen sensor network for monitoring radiation worldwide.

Using Arduino as the backbone of a DIY Geiger counter, we'll show you how to join them.

It's Never Too Soon to Start

Don't wait for an emergency to begin monitoring your local rads. Common substances, like the concrete, brick, and marble used in building construction, often emit minute, nonharmful amounts of radiation. It's important to know the highs and lows of this "background radiation" in the local environment over weeks, months, and even years, so that significant increases in that radiation level are easier to detect.

What's a Geiger Counter?

When you break it down, a Geiger counter is not a very complicated device. In fact, it has more than a few things in common with the EMI monitor described in an earlier project. Just as the EMI monitor used a wire to detect electromagnetic radiation, the radiation detector uses a wire in the middle of a gas-filled metal tube to detect the charged particles given off by radioactive atoms.

As the heart of our radiation detector, we'll be using the Sensitive Geiger Counter (sku C6979ASB (*http://www.goldmine-elec-products.com/prodinfo.asp?number=C6979ASB*)) from Electronic Goldmine. To save time, we bought our Geiger counter already assembled, but if you're feeling daring, you can save some money by getting yours in kit form and assembling it yourself.

The Soviet-era SBM-20 detection tube of our Geiger counter is filled with a low-pressure mixture of neon, bromine, and argon gases. A high-voltage transformer charges a wire inside the tube with 450 volts of electricity, and an output device alerts us when the tube detects a particle. With our model, there are two output devices: a flashing LED and a speaker to make the "click" the public has come to expect from a Geiger counter.

The outside of the metal tube is connected to the ground. When no radiation is present, the gas does not conduct any electricity from the wire to the tube's metal casing. But when ionizing radiation enters the tube, the gas molecules become ionized (i.e., they lose an electron, resulting in positively charged atoms and free electrons floating around the tube). These electrically charged particles can complete a circuit between the tube's metal casing and the internal wire, which results in a short sharp shock to the gadget's circuitry.

This shock is passed to a speaker to produce an audible click, to a lamp or LED to produce a flash of light, or to an internal counter to display the number of clicks the tube receives per minute. (Some fancy Geiger counters have all three types of output.)

When a particle is detected, the Geiger tube is momentarily charged with high-voltage electricity. The current is not very strong, so it shouldn't harm you if you accidentally touch the circuitry—but don't be a smart-aleck about it. What might not be harmful to you will almost certainly be harmful to the electronic components of this device. Shorting out the circuitry with your hand will more than likely be the end of your Geiger counter. For that reason, don't touch the Geiger counter when it is powered up.

If you've electrically connected your Geiger counter to your Arduino, and then electrically connected your Arduino to your computer or to your Internet router, an unexpected short circuit in the Geiger counter can fry it, your Arduino, your computer, and your router. Don't risk it. Make a nice wooden or plastic case for your Geiger counter, and keep your contact with the circuitry to a minimum.

Because of the danger a short circuit might cause to our equipment, we're going to forgo an electrical connection between the Arduino and Geiger counter. Instead we're going to "optoisolate" the two. Optoisolation is a process of transferring electrical signals between electronic devices using light (*optic-*) instead of electricity, to ensure that voltage spikes from one device can't damage components in the other. Since the entire point of a Geiger counter is to produce voltage spikes, we consider this a prudent precaution.

We're fortunate that the our Geiger counter has very bright flashing LED output. To read that flash, we'll be using a phototransistor from a RadioShack infrared emitter and detector package (catalog #276-0142 (*http://www.radioshack.com/product/index.jsp?productId=2049723*)). The phototransistor looks just like a clear LED (and it has a lot in common with an LED), but it is specially designed to detect light, not to emit it. We'll be using it to detect the flashing light of our Geiger counter.

Make the Gadget

Although monitoring radiation can be tricky, building this gadget is pretty straightforward. More complexity comes in when programming Arduino and uploading radiation readings to Pachube.

The Infrared Emitter

The infrared emitter is unmistakable—it looks like a funky gray LED. But the collector is essentially indistinguishable from a clear LED unless you look very carefully along the vertical axis. The collector looks dark gray when seen from above, while a clear LED looks light gray when seen from above. For this reason, we suggest you keep the emitter and detector in a special place in your parts box, apart from other components—you can go crazy trying to find the collector in a sea of clear LEDs.

Parts

1. Arduino
2. Ethernet shield
3. Breadboard
4. Geiger counter
5. IR collector (RadioShack catalog #276-0142)
6. 1k resistor (color code brown, black, red)
7. Long solid strand wire, 18–22 gauge
8. 1 package of heat-shrink tubing. We used tubing from RadioShack (catalog #278-1611 (*http://www.radioshack.com/product/index.jsp?productId=2062652*)); you can use others.

Breadboard the Circuit

Check your work on the breadboard view (Figure 9-1).

Step 1 Connect one end of a long red wire to the collector lead (the shorter lead) on the IR detector and the other into a row on the breadboard.

Step 2 Connect one end of a long black wire to the emitter lead (the longer lead) on the IR detector. Plug the other end into a different row on the breadboard.

Step 3 Connect a jumper wire from the emitter on the detector to GND on Arduino.

Step 4 Insert the 1K resistor into the same row on the breadboard as the long red wire from the detector.

Step 5 Connect a jumper wire from the 5v pin on Arduino to the resistor on the breadboard.

Step 6 Connect a jumper wire from digital pin 2 on Arduino to the place on the breadboard where the collector wire meets the 1K resistor.

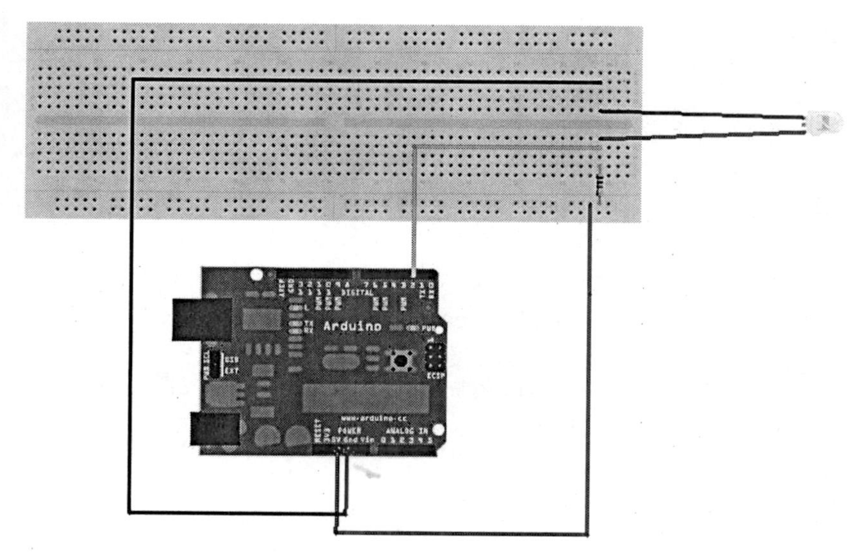

Figure 9-1. *The Geiger counter input.*

Step 7: We're going to use the output of the Geiger counter (specifically the flashing LED) as the input to our Arduino.

1. Isolate the LED with a bit of heat-shrink tubing.
2. Place the IR detector atop the Geiger counter LED. Leave a tiny gap so that they do not touch.
3. Use a blow-dryer to heat-shrink the connector.

That's it!

Write the Code

Upload the following sketch to Arduino. You can find it on EMWA GitHub repository | chapter-9 | radiation-pachube-sketch (*https://github.com/ejgertz/EMWA/blob/master/chapter-9/radiation-pachube-sketch*).

```
/*
  This code, which assumes you're using the official Arduino Ethernet
  shield, updates a Pachube feed with your analog-in values and grabs
  values from a Pachube feed--basically it enables you to have both
  "local" and "remote" sensors. Tested with Arduino 1.0.

  Pachube is www.pachube.com--connect, tag and share real time sensor data
  code by usman (www.haque.co.uk), may 2009
  copy, distribute, whatever, as you like.
```

```
  v1.1 - added User-Agent & fixed HTTP parser for new Pachube headers
  and check millis() for when it wraps around

  Ethernet shield attached to pins 10, 11, 12, 13

  http://www.tigoe.net/pcomp/code/category/arduinowiring/873
  This code is in the public domain.

  Modified autumn 2011 by Patrick Di Justo, based on code by Tom Igoe
*/

#include <SPI.h>
#include <Ethernet.h>

// assign a MAC address for the Ethernet controller.
// Newer Ethernet shields have a MAC address on a sticker on the shield
// fill in your address here:

byte mac[] = {  0xDE, 0xAD, 0xBE, 0xEF, 0xCA, 0xFE};

// initialize the library instance:
EthernetClient client;
// last time you connected to the server, in milliseconds
long lastConnectionTime = 0;
// state of the connection last time through the main loop
boolean lastConnected = false;
// delay between updates to Pachube.com
const int postingInterval = 15000;

int minuteFactor = 60000 / postingInterval;

int geiger_input = 2;
long timePreviousMeassure = 0;
long countPerMinute = 0;
long count = 0;
float radiationValue = 0.0;

// Define the SPI pin for the SD Card
#define SD_CARD 4


//This is the conversion factor for the SBM-20 radiation detection tube
#define CONV_FACTOR 0.0057

// replace the Xs with YOUR Pachube feed ID:
#define SHARE_FEED_ID XXXXX

// replace the Xs with your Pachube API key:
#define PACHUBE_API_KEY "__YOUR KEY HERE__" // fill in your API key

void setup()
```

```
{
  // If using the Wiznet SD card/Ethernet shield, these two lines
  // are absolutely necessary to temporarily disable the SD card
  // so that the Ethernet port will work.

  pinMode(SD_CARD, OUTPUT);
  digitalWrite(SD_CARD, HIGH);

  // start serial port:
  Serial.begin(9600);
  // start the Ethernet connection:
    delay(1000);

  if (Ethernet.begin(mac) == 0)
  {
    Serial.println("Failed to configure Ethernet using DHCP");
    // no point in carrying on, so do nothing forevermore:
    for(;;)
      ;
  }

  // give the Ethernet module time to boot up:
  delay(1000);

  // Set the Geiger counter input to HIGH so we can tell when it
  // changes.  We are going to use Arduino interrupt 0, connected
  // to digital pin 2, which we are using for geiger_input.

  pinMode(geiger_input, INPUT);
  digitalWrite(geiger_input,HIGH);

  attachInterrupt(0,countPulse,CHANGE);
}

void loop()
{
  if (millis()-timePreviousMeassure > postingInterval)
  {
    countPerMinute = count*minuteFactor;
    radiationValue = countPerMinute * CONV_FACTOR;
    timePreviousMeassure = millis();
    Serial.println(count);
    Serial.print("cpm = ");
    Serial.print(countPerMinute,DEC);
    Serial.print(" - ");
    Serial.print("uSv/h = ");
    Serial.println(radiationValue,4);
    count = 0;
  }

  // if there's incoming data from the net connection.
```

```
  // send it out the serial port. This is for debugging
  // purposes only:
  if (client.available()) {
    char c = client.read();
    Serial.print(c);
  }

  // if there's no net connection, but there was one last time
  // through the loop, then stop the client:
  if (!client.connected() && lastConnected) {
    Serial.println();
    Serial.println("disconnecting.");
    client.stop();
  }

  // if you're not connected, and ten seconds have passed since
  // your last connection, then connect again and send data:
  if (!client.connected() && (millis() - lastConnectionTime >
postingInterval)) {
    sendData(radiationValue);
  }

  // store the state of the connection for next time through
  // the loop:
  lastConnected = client.connected();
}

void countPulse()
{
  detachInterrupt(0);
  count++;
  digitalWrite(13,HIGH);
  while(digitalRead(2)==0){}
  digitalWrite(13,LOW);

  attachInterrupt(0,countPulse,CHANGE);
}

// this method makes a HTTP connection to the server:
void sendData(int thisData) {
  // if there's a successful connection:
  if (client.connect("www.pachube.com", 80)) {
    Serial.println("connecting...");
    // send the HTTP PUT request.
    // fill in your feed address here:
    client.print("PUT /api/");
    client.print(SHARE_FEED_ID);
    client.print(".csv HTTP/1.1\nHost: pachube.com\nX-PachubeApiKey: ");
    client.print(PACHUBE_API_KEY);
    client.print("\nContent-Length: ");
```

```
    // calculate the length of the sensor reading in bytes:
    int thisLength = getLength(thisData);
    client.println(thisLength, DEC);

    // last pieces of the HTTP PUT request:
    client.print("Content-Type: text/csv\n");
    client.println("Connection: close\n");

    // here's the actual content of the PUT request:
    client.println(thisData, DEC);

    // note the time that the connection was made:
    lastConnectionTime = millis();
  } else
  {
    // if you couldn't make a connection:
    Serial.println("connection failed");
  }
}

// This method calculates the number of digits in the
// sensor reading.  Since each digit of the ASCII decimal
// representation is a byte, the number of digits equals
// the number of bytes:

int getLength(int someValue)
{
  // there's at least one byte:
  int digits = 1;
  // continually divide the value by ten,
  // adding one to the digit count for each
  // time you divide, until you're at 0:
  int dividend = someValue /10;
  while (dividend > 0) {
    dividend = dividend /10;
    digits++;
  }

  // return the number of digits:
  return digits;
}
```

What Are We Measuring with This Gadget?

This gadget is measuring radiation in "counts per minute" (CPM), which at this writing is the most commonly used increment for sharing DIY radiation counter data on Pachube. Each time a subatomic particle ionizes the gas molecules in the detection tube, thus closing the circuit, Arduino registers that as one count.

We use CPM because we can't assume that DIY radiation detectors are calibrated to an official standard. The detector tube used in this build claims to have a factor that can convert counts per minute into sieverts, a unit of radiation measurement commonly used by scientists. But since most of us don't have access to the kinds of laboratory facilities that would allow us to confirm this calibration, counts per minute are the best units to use.

Taken over weeks, months, and years, CPM give us a useful *qualitative* measurement of radiation levels, rather than a *quantitative* measurement. That is, the readings can tell us if the radiation level changes dramatically, such as jumping from 50 to 150CPM. A significant increase like that might be worth looking into, even if we don't know exactly what it means in sieverts.

Failure Mode Analysis

This is the most complicated project in the book, so it would be amazing if your gadget worked perfectly the first time. It took us nearly a week to get all the pieces of our gadget working.

So don't be discouraged if, on your first try, your Pachube data is a big flat line of nothing.

Remember, the first point of our workbench philosophy back in Chapter 1 is to break it down when something doesn't work.

So, break it down:

- Check your build: Be sure the gadget is assembled correctly.
- Next, mentally divide the project in two parts: input (what comes into Arduino) and output (what goes out of Arduino).
  - Input:
    - Does your Geiger counter detect background radiation?
    - Does your Arduino successfully record each flash of the Geiger counter?
    - Does the data show up in the serial monitor?
    - Once you've gotten the input working, *don't fiddle with it*.
  - Output:
    - Does your Arduino show up on your local network?
    - Did you run the Arduino web page and Arduino Pachube example sketches successfully?

— Did you replace the values in the sketch with *your* IP address, gateway, subnet mask, and Pachube API code?

Troubleshoot your gadget methodically, changing only one thing at a time until you've solved a particular problem—and then simply move on to the next. Also, remember that it's okay to *ask for help*--both online on Arduino forums and at your local hacker space.

Uploading data successfully to Pachube was the hardest part of building this gadget for us. In the process we learned a valuable lesson: always use the *most recent tutorial* on the Pachube website. At the time of this writing, there are still tutorials from 2009 on the Pachube website. These are worthless. Find the most recent tutorials.

Things to Try

1. If you find yourself in possession of a few bags of high-potassium commercial fertilizer, or potassium chloride water softener tablets, or potassium chloride ice/snow melter, or potassium-based salt substitute, or even a bowl of Brazil nuts, try getting a radiation measurement from them with your Geiger counter. It's not difficult to get a reading that's twice the normal background radiation. The key is the potassium: household items that are high in potassium are *very slightly* more radioactive than other objects. This is because elemental potassium has a naturally occurring isotope called potassium-40 (about 1 in every 8,000 atoms), which is very slightly radioactive. (We keep emphasizing *very slightly* so that you don't panic at the sight of a bunch of bananas, which are high in potassium.) This very slight radioactivity is more than enough to be detected by your Geiger counter.
2. Rather than using the preassembled Geiger counter, buy a kit, such as Electronic Goldmine's Sensitive Geiger Counter Kit (sku C6979 (*http://www.goldmine-elec-products.com/prodinfo.asp?number=C6979*)), and build it yourself. Note: This kit requires intermediate-level soldering skills; we recommend it for makers who have accomplished at least a few successful soldering projects.

10/Casing the Gadget

If you want to take your Arduino gadget mobile, or simply protect it from dust, you'll need to secure it inside some sort of portable, durable case. Properly, such cases are called "Arduino project enclosures."

Since almost any smallish, box-like object has enclosure potential, this is an arena where your creativity can really take off. Just browse any retail store specializing in home, school, and office storage products, and your head will start to spin with the possibilities.

Also, don't forget to think outside the enclosure box! Hardware stores, dollar stores, craft stores, toy stores—these and more have all sorts of products that may inspire you to adapt, or to design and create, a totally original container to enclose your gadget.

You can also check out DIY websites like Instructables (*http://www.instructables.com*) and MAKE Magazine (*http://www.makezine*) for ideas and examples.

All this said, building enclosures does add time to a project, and doesn't appeal to everyone. So in the last few years, a variety of ready-made Arduino project enclosures have become available for purchase, often at the same sites that sell electronic components.

While we don't endorse any particular enclosure, here are a few suppliers and products to consider:

Adafruit Industries (*http://www.adafruit.com*): Enclosure for Arduino (ID: 271) Clear Enclosure for Arduino (ID: 337)

SparkFun Electronics: (*http://www.sparkfun.com*) Crib for Arduino (sku: PRT-10033) Arduino Project Enclosure (sku: PRT-10088)

Solarbotics: (*http://www.solarbotics.com*) Solarbotics Arduino Freeduino Enclosure (sku: 60100)

Nathan Masuda's Shapeways Shop: (*http://shpws.me/F2Z*) Stackable Arduino Enclosure

More enclosures may be on the market by the time you read this book, so be sure to look around online, or ask friends at your local hacker space what they recommend.

CPSIA information can be obtained at www.ICGtesting.com
Printed in the USA
LVOW072358210213

321236LV00003B/95/P